Characterization Problems Associated with
the Exponential Distribution

T.A. Azlarov   N.A. Volodin

Characterization Problems Associated with the Exponential Distribution

Translated by Margaret Stein and Edited by Ingram Olkin

Springer-Verlag
New York Berlin Heidelberg Tokyo

Margaret Stein (*Translator*)
Ingram Olkin (*Editor*)
Department of Statistics
Stanford University
Stanford, CA 94305
U.S.A.

AMS Classifications: 60E10, 62E99

Library of Congress Cataloging-in-Publication Data
Azlarov, T.A.
   Characterization problems associated with the exponential distribution.
   Translation of: Kharakterizatsionnye zadachi, sviazannye s pokazatel´nym raspredeleniem.
   Bibliography: p.
   Includes index.
   1. Distribution (Probability theory)   I. Volodin, N.A.
II. Olkin, Ingram.   III. Title.
QA273.6.A9613   1986   519.2   86-3786

Original edition: *Kharakterizatsionnye Zadachi, Sviazannye S Pokazatel'nym Raspredeleniem*
Taschkent. © Izdatel'stvo "Fan" Uzbekckoi CCR, 1982.

© 1986 by Springer-Verlag New York Inc.
Softcover reprint of the hardcover 1st edition 1986

9 8 7 6 5 4 3 2 1

ISBN 978-1-4612-9374-3   ISBN 978-1-4612-4956-6   (eBook)
DOI 10.1007/978-1-4612-4956-6

TABLE OF CONTENTS

Chapter

1   Introduction   1

    1.1 Characterization of the exponential distribution
       by the lack of memory property   5

2   Characterization Problems in the Class of
    Distributions with a Monotone Failure Rate   14

3   Some Properties of Order Statistics   24

4   Characterization of the Exponential Distribution
    by an Independent Function of its
    Order Statistics   31

5   Characterization of the Exponential Distribution
    by a Property of its Order Statistics   42

6   Characterizations of the Distributions by Moment
    Properties of Order Statistics   56

7   Characterization Problems of the Geometric
    Distribution   68

8   Characterization of the Exponential Distribution
    using the Geometric Distribution   79

9   Multivariate Exponential Distributions and their
    Characterizations   86

    Appendix   100

    Comments   113

    Bibliography   119
    Author Index   134

Chapter 1
INTRODUCTION

Problems of calculating the reliability of instruments and systems and the development of measures to increase efficiency and reduce operational costs confronted physicists and mathematicians at the end of the '40's and the beginning of the '50's in connection with the unreliability of electro-vacuum instruments used in aviation. Since then steadily increasing demands for the accuracy, reliability and complexity required in electronic equipment have served as a stimulus in the development of the theory of reliability. From 1950 to 1955 Epstein and Sobel [67, 68] and Davis [62], in an analysis of statistical data of the operating time of an instrument up to failure, showed that the distribution is exponential in many cases. Consequently, the exponential distribution became basic to research associated with experiments on life expectancy. Further research has shown that there are a whole series of problems in reliability theory for which the exponential distribution is inapplicable. However, it can practically always be used as a first approximation. The ease of computational work due to the nice properties of the exponential distribution (for example, the lack of memory property, see Section 1) is also a reason for its frequent use.

As a rule, data on the behavior of the failure rate function are used to test the hypothesis that a given distribution belongs to the class of exponential distributions, and order statistics are used to estimate the parameter of the exponential distribution. The practical side of these questions is sufficiently fully clarified in the book by Barlow and Proschan [6] and in a collection of articles edited by Sarhan and Greenberg [21]. The enormous role that the failure rate function and order statistics play in applied statistical problems in reliability theory is clearly shown in these books.

1

In the mid-1960's the study of properties characterizing the exponential distribution began. Ferguson [71-73], Rossberg [123-124] and a host of other statisticians and mathematicians obtained results characterizing the exponential distribution by means of independence or identical distribution properties of order statistics. It was emphasized that by using the characterizing properties of the exponential distribution one could obtain analogous characterizations of the Pareto, Weibull, extreme value distributions.

In their book, Kagan, Linnik and Rao [10] cited some characterizing properties of the exponential distribution that arise in estimating the parameters by the method of maximum likelihood (see also [141]) and in finding the distribution with maximum entropy in a class of distributions defined on a positive semi-axis with finite first moments.

Many characterizations of the exponential distribution have been published during the last ten years. In the work by Ahsanullah, Galambos, Klebanov, Kotz, Shimizu and others, new statements of characterization problems and estimates of stability in characterization problems of the exponential distribution appeared. In 1978 the first monograph, by Galambos and Kotz [79], devoted to characterization problems of the exponential and related distributions was published. Beginning in 1978 a series of papers by Shimizu appeared in which, with the aid of theorems of the Choquet-Deny type, new methods were developed for solving functional equations arising in the characterization of certain distributions, among them the exponential.

We were guided by the following considerations in writing this book. Firstly, we wanted to mention the results obtained by Tashkent mathematicians; secondly, to acquaint the readers with the new ideas and results that have been obtained recently and published in various, not always readily accessible, periodicals. Finally, there is no monograph (in the Russian language) devoted to characterizing properties of the exponential distribution. There is only a small section in [10] on this subject.

The authors of this monograph have attempted above all to clarify fully the possible formulations of characterization problems and methods of solution for the exponential distribution. In Section 1 one of the basic characterizing properties of the exponential distribution is studied—the lack of memory property. In Section 2 the class of distributions with monotone failure rate is described and estimates of their proximity to the exponential distribution are given. In Section 3 the basic properties of order statistics and their moments are discussed. These properties are used subsequently in the three following sections. In Sections 4 and 5 various characterizations of the exponential distribution by properties of independence and of identically distributed functions of order statistics are considered. In Section 6 the question of the unique recovery of a distribution function by a set of moments of order statistics is studied, and in Sections 7 and 8 we study the geometric distribution (which is a discrete analogue of the exponential distribution) and its relation to the exponential. In Section 9 various multivariate generalizations of the exponential distribution are given. Two theorems used in almost all sections of the book are cited in the Appendix.

Information referring to the history of the models considered and references to the literature are given in the comments at the end of the book. The bibliography is incomplete; only works relating to references in the text are given there. The history of the problems considered in this book is dealt with more fully in [79].

The authors express their deep thanks to A. A. Džamiraev, Š. A. Ismatullaev and L. B. Klebanov for thier helpful discussions and remarks that contributed to the improvement of the book, as well as to A. Obretenov and S. Rachev who sent the authors two of their unpublished articles, the results of which are included in this book.

We take this opportunity to express our sincere gratitude to Academician S. H. Siraždinov of the Uzbek Academy of Sciences for the attention and support shown to the authors during the preparation of this book.

1. Chararacterization of the exponential distribution by the lack of memory property.

We shall consider a service instrument whose operating time until failure is random. If the life expectancy of this instrument has an exponential distribution then the remaining operating time, under the condition that at a given moment the instrument is performing normally, has the same exponential distribution. Furthermore, if some occurrence is characterized by a complete lack of memory or "of aging" then the probability distribution of its duration must be exponential.

Let ξ be an arbitrary non-negative random variable that can be thought of as a lifetime or waiting time. Let us consider the tail of the distribution

$$\bar{F}(x) = 1 - F(x) = P\{\xi \geq x\}.$$

Clearly $\bar{F}(x)$ is "the probability that the lifetime (from the moment of birth) exceeds x." At a given age y, the event that the remaining lifetime will exceed $x + y$ can be written as $\{\xi \geq x + y\}$ and the conditional probability of this event (at a given age y) is equal to the ratio $P\{\xi \geq x + y\}/P\{\xi \geq x\}$. This is the distribution of the remaining lifetime. It coincides with the unconditional distribution of the lifetime if and only if

$$P\{\xi \geq x + y\} = P\{\xi \geq x\}P\{\xi \geq y\}, \quad x, y \geq 0.$$

This lack of memory property is sometimes called the Markov property of the exponential distribution.

Let us give a more precise mathematical form to our discussion.

Let $\mathcal{P}$ be the class of non-degenerate, non-negative random variables and let $\mathcal{E}$ be the class of random variables distributed according to the exponential law. Let ξ be a random variable with distribution function $F(x) = P\{\xi < x\}$ and $\bar{F}(x) = 1 - F(x) = P\{\xi \geq x\}$.

Theorem 1.1. If $\xi \in P$ then

$$P\{\xi \geq x + y \mid \xi \geq x\} = P\{\xi \geq y\} \tag{1.1}$$

holds for all $x \geq 0, y \geq 0$ if and only if $\xi \in \mathcal{E}$.

Proof. From (1.1) one easily finds that for all $x_j \geq 0, 1 \leq j \leq n$,

$$\bar{F}(x_1 + \cdots + x_n) = \bar{F}(x_1) \cdots \bar{F}(x_n).$$

If we take $x_j = x$ for all j then

$$\bar{F}(nx) = \bar{F}^n(x) \tag{1.2}$$

for all natural numbers n. Setting $x = 1/n$ we have

$$\bar{F}(1) = \bar{F}^n(1/n) \quad \text{or} \quad \bar{F}(1/n) = \bar{F}^{1/n}(1).$$

From this and from (1.2) with $x = 1/m$ we obtain

$$\bar{F}(n/m) = \bar{F}^{n/m}(1). \tag{1.3}$$

If $0 < \bar{F}(1) < 1$ then $\lambda = -\log \bar{F}(1)$ is finite and positive. Let us write (1.3) in the form $\bar{F}(r) = e^{-\lambda r}$ for all rational $r \geq 0$. Let x be irrational. Since the set of rational numbers is everywhere dense on the real line, there exist two sequences of rational numbers $\{r_{1n}\}$ and $\{r_{2n}\}$ such that

$$r_{1n} < x < r_{2n} \text{ and } \lim_{n \to \infty} r_{1n} = \lim_{n \to \infty} r_{2n} = x.$$

Since $\bar{F}(x)$ is a non-increasing function

$$e^{-\lambda r_{2n}} \leq \bar{F}(x) \leq e^{-\lambda r_{1n}}, \quad n \geq 1.$$

Passing to the limit as $n \longrightarrow \infty$ we get $\bar{F}(x) = e^{-\lambda x}$ for all $x \geq 0$.

If $\bar{F}(1) = 1$ or $\bar{F}(1) = 0$ then, taking into account (1.1), we would have obtained $\bar{F}(y) \equiv 1$ or $\bar{F}(y) \equiv 0$ which contradicts the condition of the theorem since we are considering a non-degenerate random variable ξ. $\|$

Let $h(x, y) = P\{\xi \geq x + y \mid \xi \geq y\} - P\{\xi \geq x\}$. Then it follows from Theorem 1.1 that $h(x, y) \equiv 0$ if and only if $\xi \in \mathcal{E}$. The stability of this property is studied in the following theorem.

Theorem 1.2. If $\xi \in \mathcal{P}$ and

$$\sup_{\substack{x \geq 0 \\ y \geq 0}} \mid h(x, y) \mid \leq \epsilon, \quad 0 \leq \epsilon < 1,$$

then

$$\lambda^{-1} \equiv \int_0^\infty \bar{F}(x) dx < \infty$$

and

$$\sup_{x \geq 0} \mid \bar{F}(x) - e^{-\lambda x} \mid \leq 2\epsilon.$$

Proof. Since the assertion is obvious for $\epsilon \geq 1/2$ we will assume that $\epsilon < 1/2$. Using the notation introduced let us write (1.1) in the form

$$\bar{F}(x + y) = \bar{F}(y)[\bar{F}(x) + h(x, y)]. \tag{1.4}$$

Let us choose some end point x_0 that satisfies the inequality $\bar{F}(x_0) \leq 1/4$. Making use of (1.4) it is not difficult to prove that for all integers $m \geq 1$ we have

$$\bar{F}(m x_0) \leq \bar{F}(x_0)[\bar{F}(x_0) + \epsilon]^{m-1}.$$

From the monotonicity of $\bar{F}(x)$ for all $x \in [m x_0, m x_0 + x_0]$ we have the inequality

$$\bar{F}(x) \leq \bar{F}(m x_0) \leq [\bar{F}(x_0) + \epsilon]^{\frac{x}{x_0} - 2},$$

7

and for all $x \geq 3x_0$

$$\bar{F}(x) \leq \frac{16}{9} a^x.$$

where $a^x = (\frac{3}{4})^{1/x_0} < 1$.

Thus, under our assumptions, all moments of the random variable ξ are finite. In particular

$$E\xi = \int_0^\infty \bar{F}(x)dx < \infty.$$

Let $E\xi = 1/\lambda$ and integrate both sides of the identity (1.4) with respect to y:

$$\int_0^\infty \bar{F}(x+y)dy = \frac{1}{\lambda}\bar{F}(x) + \frac{1}{\lambda}\theta(x), \tag{1.5}$$

where

$$\theta(x) = \lambda \int_0^\infty \bar{F}(y)h(x,y)dy.$$

From the stated conditions and notation, $\mid \theta(x) \mid \leq \epsilon$. Introducing the notation

$$\Psi(x) = \int_0^\infty \bar{F}(x+y)dy = \int_x^\infty \bar{F}(y)dy$$

and taking into account the identity $\Psi'(x) = -\bar{F}(x)$, we can rewrite (1.5) as a differential equation

$$\Psi'(x) + \lambda\Psi(x) = \theta(x).$$

The solution of this equation that satisfies the initial condition $\Psi(0) = 1/\lambda$ is

$$\Psi(x) = \frac{1}{\lambda}e^{-\lambda x} + \int_0^x e^{-\lambda(x-u)}\theta(u)du.$$

Therefore

$$\bar{F}(x) = e^{-\lambda x} + \lambda \int_0^x e^{-\lambda(x-u)}\theta(u)du - \theta(x),$$

and since $\mid \theta(x) \mid \leq \epsilon$ we have

$$\mid \bar{F}(x) - e^{-\lambda x} \mid \leq 2\epsilon$$

8

which we also needed to prove. ‖

Let us consider a series of generalizations of Theorem 1.1. One of them lies in the fact that in equality (1.1) the variables x and y are replaced by certain random variables η and ς.

Theorem 1.3. We assume that $\xi \in P$ has a continuous distribution function and D_1 and D_2 are two families of random variables such that (i) every element of D_1 and D_2 is independent of ξ; (ii) every element of D_1 is independent of every element of D_2; (iii) for each interval $[a, b), 0 \leq a < b < \infty$, there exists a random variable $\eta_i \in D_i$ such that $P\{a \leq \eta_i < b\} > 0, i = 1, 2$; (iv) either $h(x, y) \geq 0$ for all $x \geq 0, y \geq 0$ or $h(x, y) \leq 0$ for all $x \geq 0, y \geq 0$. If with probability one $h(\eta, \varsigma) = 0$ for all $\eta \in D_1, \varsigma \in D_2$, then $\xi \in \mathcal{E}$.

Proof. If $h(x, y) \equiv 0$ then the assertion follows from Theorem 1.1. Let us assume that x_1 and y_1 are two points such that $h(x_1, y_1) > 0$. Since $F(x)$ is continuous by assumption, there exist numbers $0 \leq a < b < \infty$ and $0 \leq c < d < \infty$ such that $a \leq x_1 \leq b, c \leq y_1 \leq d$ and $h(x, y) > 0$ for all $a \leq x \leq b, c \leq y \leq d$. We fix a, b, c and d and choose $\eta \in D_\infty$ and $\varsigma \in D_\mathcal{E}$ such that $P\{a \leq \eta \leq b\} > 0$ and $P\{c \leq \varsigma \leq d\} > 0$. Since $h(\eta, \varsigma) = 0$,

$$
\begin{aligned}
0 &= \int_{-\infty}^{\infty} \int_{-\infty}^{\infty} [\bar{F}(x + y) - \bar{F}(x)\bar{F}(y)] dP\{\eta < x\} dP\{\varsigma < y\} \\
&\geq \int_{a}^{b} \int_{c}^{d} [\bar{F}(x + y) - \bar{F}(x)\bar{F}(y)] dP\{\eta < x\} dP\{\varsigma < y\}.
\end{aligned}
\tag{1.6}
$$

Since $\bar{F}(x)$ is a continuous function, $h(x, y)$ is also continuous on $[a, b]$ and $[c, d]$. From Weirstrass' theorem it follows that

$$
\bar{F}(x + y) - \bar{F}(x)\bar{F}(y) \geq \epsilon > 0 \quad \text{for all} \quad x \in [a, b], y \in [c, d].
$$

Therefore, from the choice of η and ς we conclude that the right-hand side of (1.6) is positive, which is a contradiction. The theorem is proved. ‖

Theorem 1.4. If $\xi \in P$ and $h(x, y) = 0$ for all $x \geq 0$ and at least two values of $y : y_1$ and y_2

9

such that $0 < y_1 < y_2$ and y_1/y_2 is irrational, then $\xi \in \mathcal{E}$.

Proof. Let $\mathcal{A} = \{y : h(x, y) = 0$ for all $x \geq 0\}$. The proof is based on the fact that $\mathcal{A}$ is closed with respect to summation and subtraction, that is, if y_1 and y_2 are in A then

(a) $y_1 + y_2$ are in $\mathcal{A}$, in particular all $y_1, 2y_1, \ldots,$, are in $\mathcal{A}$;

(b) $y_2 - y_1$ is in $\mathcal{A}$ if $y_2 > y_1$, $\bar{F}(y_1) \neq 0$.

The proof of (a) is trivial whereas (b) follows from the fact that for any $x \geq 0$

$$\bar{F}(y_1)\bar{F}(y_2 - y_1)\bar{F}(x) = \bar{F}(y_2)\bar{F}(x) = \bar{F}(y_2 + x)$$
$$= \bar{F}(y_1 + (y_2 - y_1 + x)) = \bar{F}(y_1)\bar{F}(y_2 - y_1 + x),$$

and therefore

$$\bar{F}(y_2 - y_1)\bar{F}(x) = \bar{F}(y_2 - y_1 + x).$$

Since y_2/y_1 is assumed to be irrational, the set $\mathcal{A}$ is everywhere dense on $[0, \infty)$ and, in the same way as in Therem 1.1, we find that

$$\bar{F}(x) = e^{-\lambda x} \quad \text{for all} \quad x \in \mathcal{A}.$$

Further reasoning is exactly the same as in Theorem 1.1. ‖

Theorem 1.5. Let ξ and η be independent, non- degenerate random variables such that

$$P\{\eta = 0\} < m \equiv P\{\xi > \eta\} < 1, \tag{1.7}$$

and the random variable η has a non-lattice distribution G_0. If $h(x, \eta) = 0$ with probability one for all $x \geq 0$, then $\xi \in \mathcal{E}$.

Proof. It is not difficult to show that the condition of the theorem is equivalent to

$$\int_0^\infty [\bar{F}(x + y) - \bar{F}(x)\bar{F}(y)]dG_0(y) = 0 \quad \text{for all} \quad x \geq 0. \tag{1.8}$$

10

Since the cases $m = 0$ and $m = 1$ are excluded by condition (1.7), there exists a unique $\lambda > 0$ that satisfies

$$\int_0^\infty e^{-\lambda x} dG_0(x) = m = \int_0^\infty \bar{F}(x) dG_0(x).$$

Let us set $H(x) = \bar{F}(x)e^{\lambda x}$ and define the probability distribution $G(x)$ in the following way: $dG(x) = m^{-1}e^{-\lambda x}dG_0(x), G(x)$ and $G_0(x)$ have common sets of points of increase. Furthermore $G(x)$ and $H(x)$ satisfy the conditions of Theorem II.1 (see Appendix) with $\delta < \lambda$. Moreover, the validity of the equation

$$\int_0^\infty H(x+y)dG(y) = \int_0^\infty (1 - F(x+y))e^{\lambda x}e^{\lambda y}dG(y)$$
$$= \bar{F}(x)e^{\lambda x}m^{-1}\int_0^\infty \bar{F}(y)dG_0(y) = H(x)$$

for all $x \geq 0$ follows from (1.8). From Appendix Theorem II.1 it follows that $H(x)$ is bounded for $x \geq 0$, and for all $x \geq 0$ and points u in the set of points of increase of the function $G_0(x)$ the identity $H(x) = H(x + u)$ holds. Consequently, $H(x)$ is a constant, which proves the assertion of the theorem. $\|$

Let us introduce the function $R(x)$, defined on $[0, \infty)$ and satisfying the condition

$$\mid R(x) \mid \leq R_0 e^{-\epsilon x} \quad \text{for all} \quad x \geq 0 \tag{1.9}$$

where R_0 and ϵ are positive constants.

In the following theorem a bound for the stability of the characterizing property obtained in Theorem 1.5 is studied.

Theorem 1.6. Let ξ and η be independent, non- negative random variables that satisfy condition (1.7), and let the random variable η have a non-lattice distribution $G_0(x)$. If there exists a real function $R(x)$ satisfying condition (1.9) such that

$$P\{\xi \geq \eta + x \mid \xi \geq \eta\} = P\{\xi \geq x\}(1 - R(x)), \quad x \geq 0, \tag{1.10}$$

11

then

$$\sup_{x \geq 0} | \bar{F}(x) - e^{-\lambda x} | e^{\lambda x} \leq \delta, \quad 0 \leq \delta < 1,$$

where $R_0 \leq \frac{1-c}{4}\delta$ and λ and c are given by

$$P\{\xi > \eta\} \equiv m = \int_0^\infty e^{-\lambda x} dG_0(x),$$

$$c = m^{-1} \int_0^\infty e^{-(\lambda+\epsilon)x} dG_0(x).$$

Proof. Condition (1.10) is equivalent to

$$\int_0^\infty \bar{F}(x+y) dG_0(y) = m\bar{F}(x)(1 - R(x)), \quad x \geq 0.$$

Setting $H(x) = \bar{F}(x)e^{\lambda x}$ and $dG(x) = m^{-1}e^{-\lambda x}dG_0(x)$ we get

$$H(x) = \int_0^\infty H(x+y) dG(y) + R(x)H(x).$$

It is not difficult to show that $H(x), G(y)$ and $R(x)$ satisfy all of the assumptions of Appendix Theorem II.2, from which the assertion of the theorem follows. $\parallel$

Yet another case of the stability of the lack of memory property was studied in the work of Dimitrov, Klebanov and Rachev (1982).

Theorem 1.7. Let $\bar{F}(0) = 1$ and

$$\sup_{x,y \geq 0} | \bar{F}(x+y) - \bar{F}(x)\bar{F}(y) | = \epsilon.$$

Then

$$\epsilon/3 \leq \inf_{\lambda > 0} \sup_{x > 0} | \bar{F}(x) - e^{-\lambda x} | \leq 20\epsilon.$$

The following example shows that the order of magnitude of the bound in Theorem 1.2 cannot be made smaller than ϵ.

Example 1.1. Let

$$P\{\xi \geq x\} = \begin{cases} e^{-x}, & 0 \leq x \leq 1, \\ \frac{1}{1+\epsilon} e^{-x} & x > 1. \end{cases}$$

Then

$$\lambda = (E\xi)^{-1} = 1 + \frac{\epsilon}{e + \epsilon(e-1)},$$

and for $0 \leq x \leq 1$

$$P\{\xi \geq x\} - e^{-\lambda x} = e^{-x} - e^{-x-\epsilon x a} = e^{-x}(1 - e^{-\epsilon x a}),$$

where $a = [e + \epsilon(e-1)]^{-1}$. Hence

$$\sup_{x \geq 0} \mid P\{\xi \geq x\} - e^{-\lambda x} \mid \geq C_1 \epsilon,$$

where C_1 is an absolute constant. At the same time it can easily be shown that

$$\sup_{x,y \geq 0} \mid P\{\xi \geq x + y \mid \xi \geq x\} - P\{\xi \geq x\} \mid < \epsilon. \quad \parallel$$

Chapter 2
CHARACTERIZATION PROBLEMS IN THE CLASS OF DISTRIBUTIONS WITH A MONOTONE FAILURE RATE

Let us consider a random variable in the class $\mathcal{P}$. Let $F(x)$ be the distribution function of some random variable ξ. The function

$$R(x,y) = \frac{F(x+y) - F(x)}{1 - F(x)} \tag{2.1}$$

is called the *failure rate function*. If $F(x)$ is interpreted as the probability that some instrument is working properly during the time x, then $R(x,y)$ is the probability of failure during a finite time interval y under the condition that the instrument was working at time x. The set $\mathcal{X} \times [0,\infty)$, where $\mathcal{X} = \{x : F(x) < 1\}$ is the region over which the function $R(x,y)$ is defined.

Definition 2.1. A continuous distribution $F(x)$ is called a distribution with an increasing (decreasing) failure rate function— IFR(DFR)-distribution—if $R(x,y)$ is an increasing (decreasing) function of x for every fixed $y > 0$ in its region of definition.

If $F(x)$ has density $f(x)$ then the density of the failure rate is defined for all $x \in \mathcal{X}$ in the following manner:

$$r(x) = \frac{f(x)}{1 - F(x)}. \tag{2.2}$$

If $R(x,y)$ is divided by y and $y \longrightarrow 0$ then we obtain $r(x)$. One can show that if the distribution function $F(x)$ has density $f(x)$, then $F(x)$ is an IFR(DFR) distribution if and only if $f(x)$ is an increasing (decreasing) function.

The function $r(x)$ has the following probabilistic meaning: $r(x)dx$ is the probability that the instrument which had been working during the time x will fail in the time interval $[x, x+dx]$.

14

This function turns out to be important in many applications: in demography, in extreme value theory, in reliability theory.

The class of IFR-distributions is of great interest fundamentally since many materials and instruments in the course of time are subjected to wear—"they age", and the duration of their working properly, as a rule, has an IFR-distribution.

On the other hand, if there are several instruments to each of which there corresponds a different but constant failure rate, then the grouped intervals between failure have a DFR-distribution. This fact is treated in detail in Proschan's article in the appendix of [6].

The assumption of monotonicity of the failure rate function appears to be quite a strong limitation and in essence narrows the class of distributions being considered. However, results that are obtained for such distributions turn out to be non-trivial and can be useful in applications.

In the class of distributions with a monotone failure rate the exponential distribution has the remarkable property that it is simultaneously an IFR- and a DFR-distribution, i.e. $R(x, y)$ does not depend on x. This property characterizes exponential distributions.

Theorem 2.1. If $\xi \in \mathcal{P}$ then $F(x)$ is simultaneously an IFR- and DFR-distribution if and only if $\xi \in \mathcal{E}$.

Proof. If $\xi \in \mathcal{E}$ then a simple calculation shows that its distribution function $F(x)$ is an IFR- and DFR-distribution siimultaneously.

If $F(x)$ is simultaneously an IFR- and DFR-distribution then from (2.1) it follows that

$$\frac{F(x + y) - F(x)}{1 - F(x)} \equiv C(y) \tag{2.3}$$

15

for all $x \geq 0$. Let us write (2.3) in the form

$$\bar{F}(x) - \bar{F}(x + y) = C(y)\bar{F}(x). \qquad (2.4)$$

Setting $x = 0$ in (2.4) we get

$$C(y) = 1 - \bar{F}(y).$$

Substituting this expression for $C(y)$ in (2.4) we have

$$\bar{F}(x) - \bar{F}(x + y) = [1 - \bar{F}(y)]\bar{F}(x).$$

Removing the parentheses and cancelling like terms we get

$$\bar{F}(x + y) = \bar{F}(x)\bar{F}(y) \qquad (2.5)$$

for all $x \geq 0$ and $y \geq 0$. The solution of (2.5), as was indicated in Theorem 1.1, is $\bar{F}(x) = e^{-\lambda x}$. ‖

Theorem 2.2. If $\xi \in \mathcal{P}$ then

$$r(x) \equiv \lambda = \text{const.} \qquad (2.6)$$

holds if and only if $\xi \in \mathcal{E}$.

Proof. If $\xi \in \mathcal{E}$ then (2.6) is proved by a simple calculation.

Assume (2.6) holds; we write it in the form

$$\frac{\bar{F}'(x)}{\bar{F}(x)} = -\lambda.$$

The solution of this equation is $\bar{F}(x) = e^{-\lambda x + C}$ where $C = 0$ by virtue of the boundary condition $F(0) = 0$. ‖

Let us study the stability in the characterization problems which have been considered.

16

Theorem 2.3. Assume $\xi \in P$, $F(0) = 0$, and that there exists a function $C(y)$ such that

$$\sup_{\substack{x \geq 0 \\ y \geq 0}} \mid R(x,y) - C(y) \mid < \epsilon. \tag{2.7}$$

Then

$$\mu_1 = \int_0^\infty \bar{F}(x)dx < \infty$$

and

$$\sup_{x \geq 0} \mid \bar{F}(x) - e^{x/\mu_1} \mid < 4\epsilon.$$

Proof. From (2.7) we obtain immediately

$$\sup_{y \geq 0} \mid C(y) - F(y) \mid < \epsilon.$$

Substituting this estimate in (2.7) we have

$$\sup_{\substack{x \geq 0 \\ y \geq 0}} \mid \frac{F(x+y) - F(x)}{1 - F(x)} - F(y) \mid < 2\epsilon$$

or

$$\sup_{\substack{x \geq 0 \\ y \geq 0}} \mid \bar{F}(y) - \frac{\bar{F}(x+y)}{\bar{F}(x)} \mid < 2\epsilon.$$

Using the result of Theorem 1.2 we obtain the assertion of the theorem. ‖

Theorem 2.4. If $\xi \in P$ and

$$\sup_{x \geq 0} \mid r(x) - \lambda \mid < \lambda\epsilon \tag{2.8}$$

where λ is a positive constant and $0 \leq \epsilon < 1$, then

$$\sup_{x \geq 0} \mid \bar{F}(x) - e^{-\lambda x} \mid \leq \frac{\epsilon}{e(1 - \epsilon)}. \tag{2.9}$$

Proof. The inequality (2.8) can be written as

$$-\frac{\bar{F}'(x)}{\bar{F}(x)} = \lambda(1 + \epsilon\theta(x)), \tag{2.10}$$

17

where $\theta(x)$ is a function such that $\mid \theta(x) \mid \leq 1$. The solution of (2.10), taking into account that $F(0) = 0$, is

$$\bar{F}(x) = \exp\{-\lambda x - \lambda\epsilon \int_0^x \theta(y)dy\}.$$

Obviously

$$\bar{F}(x) - e^{-\lambda x} \leq e^{-\lambda x + \lambda\epsilon x} - e^{-\lambda x} \leq \lambda\epsilon x e^{-\lambda x(1-\epsilon)}$$

and

$$\bar{F}(x) - e^{-\lambda x} \geq -\lambda\epsilon x e^{-\lambda x}$$

from which we obtain

$$\mid \bar{F}(x) - e^{-\lambda x} \mid \leq \lambda\epsilon x e^{-\lambda x(1-\epsilon)}.$$

Maximizing the right-hand side of the last inequality we get (2.9).‖

Let us mention several properties of distributions with a monotone failure rate. Proofs of these properties can be found in the second chapter of [6].

Property 2.1. The following assertions are equivalent:

(a) $F(x)$ is an IFR (DFR)-distribution;

(b) the function $\log \bar{F}(x)$ is concave (convex) for all $x \in X$;

Property 2.2. If $\bar{F}(x)$ is an IFR (DFR)-distribution then

$$[1 - F(x)]^{1/x}$$

decreases (increases) as x increases.

We note that if $F(x)$ is an IFR-distribution, then the random variable ξ has finite moments of all orders. Therefore, the problem of estimating the closeness of an IFR—distribution to some exponential distribution by the method of moments seems completely natural. And if we

18

assume the existence of a certain number of moments of a DFR-distribution, then an analogous problem can be posed in this case also.

Let $\mu_n = \int_0^\infty x^n dF(x)$, $\quad n = 1, 2, \ldots$

Theorem 2.5. Assume that $F(x), x \geq 0$, has a monotone failure rate density. Then

$$\sup_{x \geq 0} | \bar{F}(x) - e^{x/\mu_1} | \leq \sqrt{2(1 - \mu_1 r(0))(1 - \frac{\mu_2}{2\mu_1^2})}.$$

Proof. Let us introduce an auxiliary random variable η with distribution function $\frac{1}{\mu_1} \int_0^\infty [1 - F(y)]dy$. The random variable η has the failure rate density

$$r_1(x) = \frac{1 - F(x)}{\int_x^\infty [1 - F(y)]dy}.$$

It is not difficult to show that if ξ has an IFR (DFR)-distribution, then η also has an IFR (DFR)-distribution and $r_1(x) \geq r(x)(r_1(x) \leq r(x))$.

Let us consider the function

$$\varphi(x) = 1 - F(x) - \frac{1}{\mu_1} \int_x^\infty [1 - F(y)]dy. \tag{2.11}$$

Since $\varphi(x) = [r_1(x) - r_1(0)] \int_x^\infty [1 - F(y)]dy$, $\varphi(x) \geq 0$ for all $x \geq 0$ for an IFR-distribution (correspondingly $\varphi(x) \leq 0$ for all $x \geq 0$ for a DFR-distribution).

Let

$$\Psi(x) = \int_x^\infty [1 - F(y)]dy.$$

Then (2.11) can be written as

$$\varphi(x) = -\Psi'(x) - \frac{1}{\mu_1} \Psi(x).$$

19

The general solution of this equation has the form

$$\Psi(x) = Ce^{-x/\mu_1} - \int_0^x e^{(y-x)/\mu_1}\varphi(y)dy,$$

where the constant C is determined by the boundary condition $\Psi(0) = \mu_1$. Therefore $C = \mu_1$ and

$$\Psi(x) = \mu_1 e^{-x/\mu_1} - \int_0^x e^{(y-x)/\mu_1}\varphi(y)dy. \tag{2.12}$$

Using (2.12) the equality (2.11) can be written as follows:

$$\bar{F}(x) - e^{-x/\mu_1} = \varphi(x) - \frac{1}{\mu_1}\int_0^x e^{(y-x)/\mu_1}\varphi(y)dy. \tag{2.13}$$

Since $\varphi(x) \geq 0 (\varphi(x) \leq 0)$ for an IFR (DFR)-distribution, from (2.13) we obtain the inequality

$$\left.\begin{array}{ll} \bar{F}(x) - e^{-x/\mu_1} \leq \varphi(x) & \text{for an IFR} \\ \bar{F}(x) - e^{-x/\mu_1} \geq \varphi(x) & \text{for a DFR} \end{array}\right\}. \tag{2.14}$$

Let us consider

$$\varphi'(x) = -F'(x) + \frac{1}{\mu_1}[1 - F(x)] = [1 - F(x)][\frac{1}{\mu_1} - r(x)].$$

From this we can obtain the inequality

$$\left.\begin{array}{ll} \varphi'(x) \leq \dfrac{1}{\mu_1} - r(0) & \text{for an IFR} \\ \varphi'(x) \geq \dfrac{1}{\mu_1} - r(0) & \text{for a DFR} \end{array}\right\}. \tag{2.15}$$

Furthermore, since ξ has an IFR (DFR)-distribution there exists a unique $x_0 \in [0, \infty)$ such that $\varphi(x_0) = \max_x \varphi(x)(\varphi(x_0) = \min_x \varphi(x))$. Consequently, taking into account (2.14) we obtain

$$\left.\begin{array}{ll} \bar{F}(x) - e^{-x/\mu_1} \leq \varphi(x_0) & \text{for an IFR} \\ \bar{F}(x) - e^{-x/\mu_1} \geq \varphi(x_0) & \text{for a DFR} \end{array}\right\}. \tag{2.16}$$

On the other hand it follows from (2.13) that

$$
\begin{aligned}
\bar{F}(x) - e^{-x/\mu_1} &\geq -\frac{1}{\mu_1}\int_0^x e^{(y-x)/\mu_1}\varphi(y)dy \geq -\varphi(x_0)\frac{1}{\mu_1}\int_0^x e^{(y-x)/\mu_1}dy \geq -\varphi(x_0) \quad \text{for an IFR} \\
\bar{F}(x) - e^{-x/\mu_1} &\leq -\varphi(x_0) \qquad\qquad\qquad\qquad\qquad\qquad\qquad\qquad\qquad\qquad\quad \text{for a DFR}
\end{aligned}\Bigg\} .
$$

$$(2.17)$$

Inequalities (2.16) and (2.17) yield

$$
\sup_x \mid \bar{F}(x) - e^{-x/\mu_1} \mid \leq \mid \varphi(x_0) \mid .
\tag{2.18}
$$

Let us consider in further detail only the case of an IFR-distribution (the reasoning for a DFR-distribution differs only by a change in the signs of the inequality).

Let $x \leq x_0$. Integrating (2.15) from x to x_0 we have

$$
\varphi(x_0) - \varphi(x) \leq \frac{x_0 - x}{\mu_1}(1 - \mu_1 r(0)).
$$

Taking into account that $\varphi(0) = 0$ we get $\varphi(x_0) \leq \frac{x_0}{\mu_1}(1 - \mu_1 r(0))$. Since $\varphi'(x)$ decreases monotonically on $[0, x_0]$ the function $\varphi(x)$ is convex upwards on this interval. Therefore

$$
\begin{aligned}
\mu_1 - \frac{\mu_2}{2\mu_1} = \int_0^\infty \varphi(x)dx &\geq \int_0^{x_0} \varphi(x)dx \geq \frac{1}{2}x_0\varphi(x_0) \\
&\geq \frac{\mu_1}{2}\varphi^2(x_0)(1 - \mu_1 r(0))^{-1}.
\end{aligned}
$$

Substituting the bound

$$
\varphi(x_0) \leq \sqrt{2(1 - \mu_1 r(0))(1 - \frac{\mu_2}{2\mu_1^2})}
$$

for $\varphi(x_0)$ in (2.18), we obtain the assertion of the theorem. ‖

Corollary 2.1. If $F(x)$ is a distribution with a monotone failure rate density then

$$
\sup_x \mid \bar{F}(x) - e^{-x/\mu_1} \mid \leq \mid 1 - \mu_1 r(0)) \mid .
$$

21

Proof. Let us prove the assertion for an IFR-distribution Everything is done analogously for a DFR-distribution.

Let x_0 be the point of maximum of the function $\varphi(x)$. Since

$$\varphi'(x) \leq [1 - F(x)][\frac{1}{\mu_1} - r(0)],$$

integrating this inequality from 0 to x_0 we obtain

$$\varphi(x_0) - \varphi(0) \leq [\frac{1}{\mu_1} - r(0)] \int_0^\infty [1 - F(x)]dx = 1 - \mu_1 r(0)$$

which together with (2.18) proves the assertion since $\varphi(0) = 0$. $\|$

In the following example we shall show that the bounds obtained in Theorem 2.5 and Corollary 2.1 cannot be improved within the accuracy of a constant.

Example 2.1. Let us consider a random variable ξ with density

$$f(x) = \begin{cases} e^{-x}(1 - \epsilon + \epsilon x) & \text{if } x \geq 0 \\ 0 & \text{if } x < 0. \end{cases}$$

It is not difficult to verify that $\bar{F}(x) = (1 + \epsilon x)e^{-x}$, $\mu_1 = 1 + \epsilon$, $\mu_2 = 2(1 + \epsilon)$ in this case. It can be shown by a direct computation that

$$\sup_x | \bar{F}(x) - e^{-x/(1+\epsilon)} | \geq | \bar{F}(1) - e^{-1/(1+\epsilon)} | \geq \frac{\epsilon^2}{8e}$$

for $\epsilon < 1/4$. For this case the bounds obtained in Theorem 2.5 and Corollary 2.1 are $\frac{\epsilon^2}{1+\epsilon}$ and ϵ^2 respectively. $\|$

Corollary 2.2. Let $F(x)$ be a distribution function with a monotone failure rate density. The relation $\mu_2 = 2\mu_1^2$ holds if and only if $\bar{F}(x) = \exp\{-x/\mu_1\}$.

Corollary 2.3. Let the distribution $F(x)$ have a monotone failure rate density. The relation $f(0) \equiv r(0) = \frac{1}{\mu_1}$ holds if and only if $\bar{F}(x) = \exp\{-x/\mu_1\}$.

22

It is well known (see [40]) that the moments $\{\mu_n\}$ of an IFR- distribution satisfy the inequality:

$$\frac{\mu_{n+1}}{(n+1)!}\frac{\mu_{n-1}}{(n-1)!} \le (\frac{\mu_n}{n!})^2, \quad n \ge 1. \tag{2.19}$$

Therefore if we define $\alpha_n = \mu_{n+1}/\mu_n(n+1)$, then from relation (2.19) it follows that $\alpha_n \ge \alpha_{n+1}$ for all n. Since $\alpha_n \ge 0, \alpha = \lim\limits_{n \to \infty} \alpha_n$ exists.

We shall give without proof another theorem, proved by A. Obretenov [19].

Theorem 2.6. Let $F(x)$ be an IFR-distribution. Then

$$\sup_x \mid \bar{F}(x) - e^{-x/\mu_1} \mid \le 1 - \frac{\alpha}{\mu_1}.$$

It can be shown that $\alpha_n = 1 + \frac{\epsilon}{1+\epsilon n}$ and $\alpha = 1$ in Example 2.1. Therefore the bound in Theorem (2.6) for this example is only of order $\epsilon/(1+\epsilon)$.

23

Chapter 3

SOME PROPERTIES OF ORDER STATISTICS

Let $\xi_1, \xi_2, \ldots, \xi_n$ be independent, identically distributed random variables with common distribution function $F(x)$. Let us consider the order statistics $\xi_{1,n} \leq \xi_{2,n} \leq \cdots \leq \xi_{n,n}$ corresponding to the random variables $\xi_1, \ldots, \xi_n$.

The distribution of the kth order statistic $\xi_{k,n}$ from a sample of size n will be denoted by $F_{k,n}(x)$. It is well known (see for example, [9]) that

$$F_{k,n}(x) = P\{\xi_{k,n} < x\} = \sum_{i=k}^{n} \binom{n}{i} F^i(x)[1 - F(x)]^{n-i} = k\binom{n}{k} \int_0^{F(x)} t^{k-1}(1-t)^{n-k}dt.$$

The density of the distribution of $\xi_{k,n}$ is

$$f_{k,n}(x) = k\binom{n}{k} F^{k-1}(x)[1 - F(x)]^{n-k} f(x)$$

where $f(x) = F'(x)$.

The joint density of the distribution of several order statistics $\xi_{r_1,n}, \xi_{r_2,n}, \ldots, \xi_{r_k,n}$ $(1 \leq r_1 < r_2 < \ldots < r_k \leq n; \ 1 \leq k \leq n)$ for $x_1 \leq x_2 \leq \ldots \leq x_k$ has the form

$$f_{r_1, r_2, \ldots, r_k}(x_1, x_2, \ldots, x_k) = n![\prod_{i=1}^{k} f(x_i)] \prod_{i=0}^{k} \left\{ \frac{[F(x_{i+1}) - F(x_i)]^{r_{i+1} - r_i - 1}}{(r_{i+1} - r_i - 1)!} \right\}, \tag{3.1}$$

where $x_0 = 0, x_{k+1} = +\infty, r_0 = 0, r_{k+1} = n + 1$.

The difference between consecutive order statistics is called the *spacing* and is denoted by $d_{r,n} = \xi_{r+1,n} - \xi_{r,n}$, $r = 1, \ldots, n-1$. We shall also consider spacings normalized by the factor $(n - r)$ and shall denote these by $D_{r,n} = (n - r)d_{r,n}$.

24

The density of the distribution of the r-th spacing is

$$f_{d_{r,n}}(x) = \frac{n!}{(r-1)!(n-r-1)!} \int_{-\infty}^{\infty} F^{r-1}(y)[1 - F(x+y)]^{n-r-1} f(y)f(x+y)dy. \qquad (3.2)$$

When $f(x) = \lambda e^{-\lambda x}$, $\lambda > 0$, that is, the random variables $\xi_1, \xi_2, \ldots, \xi_n$ have an exponential distribution, the density of the distribution of the r-th spacing becomes quite simple:

$$f_{d_{r,n}}(x) = (n-r)\lambda e^{-\lambda(n-r)x}, \quad x \geq 0, \qquad (3.3)$$

so that $f_{D_{r,n}}(x) = \lambda e^{-\lambda x}$. Hence it is evident that if a sample is taken from an exponential distribution, all the $D_{r,n}$, $r = 1, \ldots, n-1$ are distributed in the same way as well. Furthermore, in Section 5 we show that under certain restrictions on the class of distributions, this property characterizes the exponential distribution. Let us look at yet another property of spacings. Note that for $k = n$ formula (3.1) becomes

$$f_{1,2,\ldots,n}(x_1, \ldots, x_n) = n! \prod_{i=1}^{n} f(x_i).$$

In this case, for the exponential distribution we obtain

$$f_{1,2,\ldots,n}(x_1, \ldots, x_n) = n! \exp\{-\sum_{i=1}^{n} x_i\}, \quad 0 \leq x_1 \leq \ldots \leq x_n < \infty.$$

The last expression can be written as

$$n! \exp\{-\sum_{i=1}^{n}(n-i+1)(x_i - x_{i-1})\}, \quad \text{where } x_0 = 0.$$

If we set $y_i = (n-i+1)(x_i - x_{i-1})$ then it is clear that the variables $D_{i,n}$, $i = 0, 1, \ldots, n-1$, are distributed on the interval $(0, \infty)$ and are statistically independent random variables. These relations allow us to express $\xi_{i,n}$ in the form

$$\xi_{i,n} = \sum_{k=0}^{i-1} D_{k,n}/(n-k), \qquad (3.4)$$

25

that is, as a linear function of exponentially distributed independent random variables.

Theorem 3.1. If $\xi_{1,r} \leq \xi_{2,n} \leq \ldots \leq \xi_{n,n}$ are the order statistics for a sample from a continuous distribution with a strictly increasing distribution function $F(x)$, then the random variables $\{\xi_{1,n}, \xi_{2,n}, \ldots, \xi_{n,n}\}$ form a Markov chain.

Proof. The transformation $y = F(x)$ transforms the $\xi_{r,n}$ into $\eta_{r,n}, r = 1, \ldots, n$—order statistics from a uniform distribution on $[0, 1]$. Since the variable $z = -\log y$ is a decreasing function of y and $-\log \eta_{r,n}$ has an exponential distribution with $\lambda = 1$, the $\varsigma_{r,n}$ defined by

$$\varsigma_{r,n} = -\log \eta_{n-r+1,n}, \quad r = 1, 2, \ldots, n$$

are order statistics. Therefore, taking into account (3.4), the $\xi_{n-r+1,n}$ can be expressed in the form

$$\xi_{n-r+1,n} = F^{-1}\{\eta_{n-r+1,n}\} = F^{-1}\{e^{-\varsigma_{r,n}}\}$$
$$= F^{-1}\{\exp[-(\frac{D_{0,n}}{n} + \frac{D_{1,n}}{n-1} + \ldots + \frac{D_{r-1,n}}{n-r+1})]\}. \tag{3.5}$$

Now we can write

$$\xi_{n-r,n} = F^{-1}\{\exp[\log F(\xi_{n-r+1,n}) - \frac{D_{r,n}}{n-r}]\}$$

from which, because of the independence of $\xi_{n-r+1,n}$ and $D_{r,n}$ and equality (3.5), it follows that $\xi_{n,n}, \xi_{n-1,n}, \ldots, \xi_{1,n}$ form a Markov chain. The variables $\xi_{1,n}, \xi_{2,n}, \ldots, \xi_{n,n}$ also have this property.$\|$

Corollary 3.1. For a random sample of size n from a continuous distribution, the conditional distribution of the variable $\xi_{s,n}$, under the condition $\xi_{r,n} = x_r (s > r)$, coincides with the distribution of the $(s-r)$-th order statistic in a sample size of $n - r$ from this same distribution truncated from the left at the point $x = x_r$.

Let us look at some other characteristics encountered in the theory of order statistics.

Let $\{\xi_k\}_{k=1}^{\infty}$ be a sequence of independent, identically distributed random variables with continuous distribution function $F(x)$. The random variable ξ_j is the record value of this sequence if $\xi_j > \max\{\xi_1, \ldots, \xi_{j-1}\}$. By agreement ξ_1 is also a record value. We define $L(0) \equiv 1, L(n) = \min\{j : j > L(n-1), \xi_j > \xi_{L(n-1)}\}$, $n > 1$. Then $\{\xi_{L(k)}, k \geq 0\}$ will be a sequence of record values. Let us introduce the notation $\varsigma_n = \xi_{L(n)} - \xi_{L(n-1)}$. Analogous to the procedure for order statistics, it can be shown that $\{\xi_{L(k)}\}_{k \geq 1}$ is a Markov chain and

$$P\{\xi_{L(k)} \geq x \mid \xi_{L(k-1)} = y\} = \begin{cases} \bar{F}(x)/\bar{F}(y), & x \geq y, \\ 0, & x < y. \end{cases} \tag{3.6}$$

If we denote $R(y) = -\log \bar{F}(y)$ and $r(y) = R'(y)$ then the density of the joint distribution of the random variables $\xi_{L(0)}, \xi_{L(1)}, \ldots, \xi_{L(n)}$ is found by the formula

$$f(x_0, x_1, \ldots, x_n) = \begin{cases} r(x_0)r(x_1)\cdots r(x_n), & \text{for } -\infty < x_0 < x_1 < \ldots < x_n < \infty \\ 0, & \text{otherwise.} \end{cases} \tag{3.7}$$

By integration of formula (3.7) it is easy to derive the density of the joint distribution of $\xi_{L(n-1)}$ and $\xi_{L(n)}$:

$$f(x_{n-1}, x_n) = \begin{cases} R^{n-1}(x_{n-1})r(x_{n-1})f(x_n)/\Gamma(n), & -\infty < x_{n-1} < x_n < \infty \\ 0, & \text{otherwise.} \end{cases}$$

Hence one can compute the density of the joint distribution of ς_n and $\xi_{L(n-1)}$:

$$f_{\varsigma_n, \xi_{L(n-1)}}(z, u) = \begin{cases} R^{n-1}(u)r(u)f(u+z)/\Gamma(n), & -\infty < u < z < \infty, \\ 0, & \text{otherwise,} \end{cases} \tag{3.8}$$

and the distribution of the random variable $\xi_{L(n)}$:

$$P\{\xi_{L(n)} < x\} = \int_{-\infty}^{x} R^n(y)dF(y)/\Gamma(n). \tag{3.9}$$

Let us now give some definitions and properties of moments of order statistics.

If $\mu_{k,n}^{(m)}$ is the m-th moment of the random variable $\xi_{k,n}$, then

$$\mu_{k,n}^{(m)} = \int_{-\infty}^{\infty} x^m dF_{k,n}(x) = E\xi_{k,n}^m. \tag{3.10}$$

From the representation of $F_{k,n}(x)$ and formula (3.10) it follows that

$$\mid \mu_{k,n}^{(m)} \mid \leq k\binom{n}{k} E \mid \xi_1 \mid^m \tag{3.11}$$

and

$$\mu_{k,n}^{(m)} = k\binom{n}{k} \int_0^1 [F^{-1}(x)]^m x^{k-1} (1-x)^{n-k} dx, \tag{3.12}$$

where $F^{-1}(x) = \inf\{t : F(t) \geq x\}$. It follows from inequality (3.11) that, with $E \mid \xi_1 \mid^m < \infty$, all moments $\mu_{k,n}^{(m)}$ exist and can be computed using (3.12).

However, the finiteness of $E \mid \xi_1 \mid^m$ is only one of the sufficient conditions for the finiteness of moments of order statistics. Another sufficient condition is given by the following theorem.

Theorem 3.2. If $E \mid \xi_1 \mid^\delta < \infty$ for some $\delta > 0$, then $E \mid \xi_{k,n} \mid^m < \infty$ for all n and k that satisfy the inequality

$$m\delta^{-1} \leq k \leq n + 1 - m\delta^{-1}. \tag{3.13}$$

Proof. Let $\bar{F}(x) = 1 - F(x)$ and set $E \mid \xi_1 \mid^\delta = C$. By Chebychev's inequality $x^\delta \bar{F}(x) \leq C$ for all $x > 0$. Integrating by parts we get

$$C = -\int_{-\infty}^{\infty} \mid x \mid^\delta \, d\bar{F}(x) \geq -\int_0^A x^\delta d\bar{F}(x) = -A^\delta \bar{F}(A) + \delta \int_0^A \bar{F}(x) x^{\delta-1} dx$$

$$\geq -C + \frac{\delta}{C^{p-1}} \int_0^A [\bar{F}(x) x^\delta]^p \frac{dx}{x} \geq -C - \frac{\delta}{\delta p C^{p-1}} \int_0^A x^{p\delta} d\bar{F}^p(x).$$

Since $A > 0$ is arbitrary

$$-\int_0^{\infty} x^{p\delta} d\bar{F}^p(x) = \lim_{A \to \infty} [-\int_0^A x^{p\delta} d\bar{F}^p(x)] \leq 2pC^p. \tag{3.14}$$

28

If $\xi_{k,n}$ is the k-th order statistic then [see [9], p. 17]

$$E\{\xi_{k,n}^m; \xi_{k,n} \geq 0\} \leq -B^{-1}(k,n) \int_0^\infty x^m d\bar{F}^{n-k+1}(x), \qquad (3.15)$$

where $B(k,n) = \int_0^1 t^{k-1}(1-t)^{n-k} dt$. From (3.14) it follows that the right-hand side of (3.15) is bounded if $m \leq \delta(n-k+1)$ or $k \leq n+1-m\delta^{-1}$ which is satisfied by virtue of condition (3.13).

Completely analogously one can obtain the bounds

$$\int_{-\infty}^0 \mid x \mid^{p\delta} dF^p(x) \leq 2pC^p$$

and

$$E\{\mid \xi_{k,n} \mid^m; \xi_{k,n} < 0\} \leq B^{-1}(k,n) \int_{-\infty}^0 \mid x \mid^m dF^k(x). \qquad (3.16)$$

The right-hand side of the bound (3.16) is finite if $m \leq \delta k$ or $m\delta^{-1} \leq k$ which is satisfied by virtue of (3.13). ‖

For $0 < k < n, n \geq 2$, we have the recurrence relation

$$(n-k)\mu_{k,n}^{(m)} + k\mu_{k+1,n}^{(m)} = n\mu_{k,n-1}^{(m)}. \qquad (3.17)$$

We shall use this formula in what follows. It is obtained from (3.12) using the following elementary representation:

$$(n-k)\binom{n}{k}(1-u) + (k+1)\binom{n}{k+1}u = n\binom{n-1}{k}.$$

Let us define

$$\mu_{i,j,n}^{(\ell,m)} = E(\xi_{i,n}^\ell \xi_{j,n}^m)$$

$$= \frac{n!}{(i-1)!(j-i-1)!(n-j)!} \int_x \int_y x^\ell y^m F^{i-1}(x)\{F(y)-F(x)\}^{j-i-1}\bar{F}^{n-j}(y)dF(x)dF(y)$$

29

and

$$\sigma_{i,j,n} = \operatorname{cov}(\xi_{i,n}, \xi_{j,n}) = \mu_{i,j,n}^{(1,1)} - \mu_{i,n}^{(1)}\mu_{j,n}^{(1)}.$$

Let us now mention some properties of moments of order statistics for the exponential distribution. Let $\bar{F}(x) = e^{-x}$. Then from formula (3.4) it follows that

$$E\xi_{r,n} = \sum_{i=1}^{r} \frac{1}{n-i-1}; \quad D\xi_{r,n} = \sum_{i=1}^{r} \frac{1}{(n-i+1)^2}.$$

Moments of higher order are also easily defined by formula (3.4), taking into account that, for the exponential distribution, the random variables $D_{i,n}$ are independent and identically distributed as ξ_1.

Chapter 4

CHARACTERIZATION OF THE EXPONENTIAL DISTRIBUTION BY AN INDEPENDENT FUNCTION OF ITS ORDER STATISTICS

In Section 3 it was shown that if a sample is taken from an exponential distribution, then the spacings $d_{r,n}$ constructed from this sample are independent random variables.

We shall show that this property characterizes the exponential distribution.

Theorem 4.1. If $\xi \in P$ and $F(x)$ is strictly increasing for all $x \geq 0$, then $\xi_{2,2} - \xi_{1,2}$ and $\xi_{1,2}$ are independent if and only if $\xi \in \mathcal{E}$.

Proof. It was shown in Section 3 that if $\xi \in \mathcal{E}$ then $\xi_{1,2}$ and $\xi_{2,2} - \xi_{1,2}$ are independent.

Let us prove the converse assertion. Let $\xi_{1,2}$ and $\xi_{2,2} - \xi_{1,2}$ be independent. Then

$$P\{\xi_{2,2} - \xi_{1,2} < x \mid \xi_{1,2} = y\} = P\{\xi_{2,2} - \xi_{1,2} < x\}$$

for almost all y. On the other hand, using Corollary 3.1 we get

$$P\{\xi_{2,2} - \xi_{1,2} < x \mid \xi_{1,2} = y\} = P\{\xi_{2,2} < x + y \mid \xi_{1,2} = y\}$$
$$= P\{\tilde{\xi}_{1,1} < x + y\} \tag{4.1}$$

where $\tilde{\xi}_{1,1}$ is a random variable with distribution function

$$P\{\tilde{\xi}_{1,1} < x\} = \begin{cases} \frac{F(x) - F(y)}{1 - F(y)}, & x \geq y, \\ 0, & \text{otherwise.} \end{cases} \tag{4.2}$$

Since the right-hand side of (4.1) does not depend on y, taking into account (4.2) we obtain

$$\frac{F(x + y) - F(y)}{1 - F(y)} = F(x),$$

31

or

$$\bar{F}(x+y) = \bar{F}(x)\bar{F}(y).$$

The theorem follows from this equation and Theorem 1.1.‖

The attempt to generalize this problem requires the solution of more complicated functional equations.

Theorem 4.2. Let $g(x, y)$ be a non-negative, measurable function and $\xi_{1,2}, \xi_{2,2}$ be the order statistics of an independent sample from an absolutely continuous distribution function $F(x)$. If $Eg(\xi_{1,2}, \xi_{2,2}) < \infty$ and the random variables $\xi_{1,2}$ and $g(\xi_{1,2}, \xi_{2,2})$ are independent, then there exists a constant λ such that

$$\int_{y}^{\infty} g(y, x)F'(x)dx = \lambda[1 - F(x)] \tag{4.3}$$

for all y.

Proof. Let $\varphi_1(t), \varphi_2(u)$ and $\varphi(t, u)$ denote the characteristic functions of the random variables $\xi_{1,2}, g(\xi_{1,2}, \xi_{2,2})$ and of the random vector $(\xi_{1,2}, g(\xi_{1,2}, \xi_{2,2}))$ respectively. Then the independence of $\xi_{1,2}$ and $g(\xi_{1,2}, \xi_{2,2})$ is equivalent to

$$\varphi(t, u) = \varphi_1(t)\varphi_2(u) \tag{4.4}$$

for all real t and u. Furthermore, the joint density of $(\xi_{1,2}, \xi_{2,2})$ is $2f(x)f(y)$ for $x < y$, and 0 otherwise. Therefore

$$\varphi(t, u) = 2 \int_{-\infty}^{\infty} \int_{x}^{\infty} \exp\{itx + iug(x, y)\}f(x)f(y)dxdy.$$

Substituting $x + z = y$ we get

$$\varphi(t, u) = 2 \int_{-\infty}^{\infty} \int_{0}^{\infty} \exp\{itx + iug(x, x + z)\}f(x)f(x + z)dxdz.$$

Analogous computations yield

$$\varphi_1(t) = 2 \int_{-\infty}^{\infty} \int_0^{\infty} e^{itx} f(x) f(x+y) dy dx$$

and

$$\varphi_2(u) = 2 \int_{-\infty}^{\infty} \int_0^{\infty} \exp\{iug(x, x+y)\} f(x) f(x+y) dy dx.$$

If we substitute the last three expressions in (4.4), then differentiate with respect to u and set $u = 0$, we obtain

$$\int_{-\infty}^{\infty} \int_0^{\infty} e^{itx} g(x, x+y) f(x) f(x+y) dy dx = \lambda \int_{-\infty}^{\infty} \int_0^{\infty} e^{itx} f(x) f(x+y) dy dx.$$

Since $g(x, x+y) \geq 0$ by assumption, using the inversion formula for characteristic functions we obtain

$$\int_0^{\infty} g(x, x+z) f(x+z) dz = \lambda \int_0^{\infty} f(x+z) dz.$$

From this it is easy to transform to the functional equation (4.3), which completes the proof.‖

Note that if $g(x, y) = G(y - x)$ for $y > x$ and 0 otherwise, then the constancy of the conditional moments

$$E\{G(\xi - x) \mid \xi \geq x\} \equiv \lambda = \text{constant} \tag{4.5}$$

for all $x \geq 0$ leads to the same functional equation

$$\int_x^{\infty} G(y - x) dF(y) = \lambda[1 - F(x)] \tag{4.6}$$

which we obtained in Theorem 4.2, and without assuming absolute continuity of the distribution function $F(x)$.

Equation (4.6) came initially from equation (4.5) by considering as $G(y - x)$ the function $(y - x)^k, k = 1, 2, \dots$. In these special cases property (4.5) turns out to be a characterization of the exponential distribution.

33

The general theory for solving equation (4.6) was studied in a series of works by R. Shimizu. He showed that, under trivial restrictions on the function $G(\cdot)$, only the exponential distribution is a solution of (4.6).

Let $\xi \in \mathcal{P}$ and let $G(x)$ be a monotone increasing function that satisfies the following conditions:

(i) $G(-0) = G(+0) = 0$;

(ii) $\mu = EG(\xi)$ exists and is positive;

(iii) there exists a positive number ϵ such that

$$\mu < \int_0^\infty e^{-\epsilon x} dG(x) < \infty. \tag{4.7}$$

Theorem 4.3. Assume that the random variable ξ and the function $G(\cdot)$ satisfy the conditions given above. Let Ω be the set of all points of increase of the non-decreasing function $G(x)$. If

$$E\{G(\xi - x) \mid \xi \geq x\} = EG(\xi) \tag{4.8}$$

for all $x \geq 0$, then the function $F(x)$ can have the form:

$$F(x) = 1 - H(x)e^{-\lambda x}, \quad x \geq 0 \tag{4.9}$$

where λ is a positive constant and $H(x)$ is a periodic function with period d for all $d \in \Omega$.

In particular, $\xi \in \mathcal{E}$ if $G(x)$ is not concentrated on the lattice of points $0, \rho, 2\rho, \ldots$ for some $\rho > 0$.

Proof. First we shall show that

$$\mu(x) \equiv \int_x^\infty G(y - x) dF(y) = \int_0^\infty [1 - F(x + y)] dG(y) \leq \mu(0) \equiv \mu \tag{4.10}$$

for all $x \geq 0$.

34

Indeed, if $A > x$ then integration by parts yields

$$\int_x^A G(y-x)dF(y) = \int_0^{A-x} G(y)dF(y+x)$$

$$= -(1-F(A))G(A-x) + \int_0^{A-x}[1-F(x+y)]dG(y).$$

Since $\mu = \mu(0) = \int_0^\infty G(y)dF(y)$ exists,

$$0 \le [1-F(A)]G(A-x) \le [1-F(A)]G(A)$$

$$\le \int_A^\infty G(y)dF(y) \longrightarrow 0 \quad \text{as} \quad A \longrightarrow \infty.$$

Hence it follows that

$$\mu(x) = \lim_{A\longrightarrow\infty} \int_x^A G(y-x)dF(y) = \int_0^\infty [1-F(x+y)]dG(y).$$

Inequality (4.10) follows from the monotonicity of $F(x)$.

The distribution of the random variable ξ under the condition that $\xi \ge x$ is

$$F_x(y) = \frac{F(y)-F(x)}{1-F(x)}, \quad y > x.$$

Then, taking into account (4.10), we obtain

$$E\{G(\xi-x) \mid \xi \ge x\} = \int_x^\infty G(y-x)d_y F_x(y)$$

$$= \frac{1}{1-F(x)} \int_x^\infty G(y-x)dF(y) = \frac{1}{1-F(x)} \int_0^\infty [1-F(x+y)]dG(y),$$

and from condition (4.8)

$$\int_0^\infty [1-F(x+y)]dG(y) = \mu[1-F(x)], \quad x \ge 0. \tag{4.11}$$

On the other hand, taking into account restriction (iii) on the function $G(x)$, one can find a unique number λ such that $\lambda > \epsilon$ and

$$\int_0^\infty e^{-\lambda x}dG(x) = \mu.$$

Let the distribution function $G_0(x)$ be defined as follows: $dG_0(x) = \mu^{-1}e^{-\lambda x}dG(x)$. We introduce $H(x) = [1 - F(x)]e^{\lambda x}$. Then (4.11) can be rewritten thus:

$$H(x) = \int_0^\infty H(x+y)dG_0(y), \quad x \geq 0.$$

Taking into account the fact that the sets of points of increase of the functions $G(x)$ and $G_0(x)$ coincide, the assertion of the theorem follows immediately from Theorem II.1. ‖

From Theorem 4.3 we get the following corollary.

Corollary 4.1. Assume that $\xi \in P$ and $E\xi^\alpha < \infty$ for some $\alpha > 0$. Then

$$\{E(\xi - x)^\alpha \mid \xi \geq x\} = E\xi^\alpha$$

holds for all $x \geq 0$ if and only if $\xi \in \mathcal{E}$.

The following theorem yields a bound for the stability of the characterization obtained in Theorem 4.3.

Theorem 4.4. Assume that $\xi \in P$ and that $G(x)$ satisfies conditions (i)-(iii). If

$$\mid E\{G(\xi - y) \mid \xi \geq y\} - a \mid \leq R(x),$$

where $R(x)$ is a real function such that $R(x) \leq R_0 e^{\epsilon x}$, then

$$\sup_x \mid \bar{F}(x) - e^{-\lambda x} \mid e^{\lambda x} \leq \frac{2}{1 - c_0}R_0,$$

where

$$c_0 = \int_0^\infty e^{-x(\lambda + \epsilon)}dG(x) \Big/ \int_0^\infty e^{-\lambda x}dG(x).$$

Proof. Let us make the same transition from the functions $F(x)$ and $G(x)$ to the functions $H(x)$ and $G_0(x)$ that was made for the proof of Theorem 4.3. We arrive at the following functional equation:

$$H(x) = \int_0^\infty H(x+y)dG_0(y) + R(x)h(x), \quad x \geq 0. \tag{4.12}$$

Since $H(x)$ is positive, we can conclude from Theorem II.2 that it is bounded. Applying Theorem II.2 it is not difficult to show that $H(x)$ can be given in the form

$$H(x) = k + \ell(x)e^{-\epsilon x}, \quad x \geq 0.$$

Substituting this expression in (4.12) we get

$$\ell(x)e^{-\epsilon x} = e^{-\epsilon x}\int_0^\infty \ell(x+y)e^{-\epsilon y}dG_0(y) + R(x)(k + \ell(x)e^{-\epsilon x}), \quad x \geq 0.$$

Hence it follows that

$$\mid \ell(x) \mid \leq c_0 \sup_{x \geq 0} \mid \ell(x) \mid + R_0 k + \frac{1-c_0}{4}\sup_{x \geq 0} \mid \ell(x) \mid, \quad x \geq 0$$

or

$$\sup_{x \geq 0} \mid \ell(x) \mid \leq \frac{4}{3(1-c_0)}R_0 k,$$

from which the assertion of the theorem is obtained. ‖

L. B. Klebanov and O. Januškjavichene [15] obtained the following result under weaker conditions than those of Theorem 4.4.

Theorem 4.5. Let $\xi \in P$ and $\mu_k < \infty$. If

$$\sup_{x \geq 0} \mid E\{(\xi - x)^k \mid \xi \geq x\} - \mu_k \mid P\{\xi \geq x\} \leq \epsilon,$$

then there exists an exponential distribution with parameter λ and positive constants C_1, C_2, ϵ_0 and $\alpha < 1$ such that for all $\epsilon \in [0, \epsilon_0]$ the following inequality holds:

$$C_1\epsilon^\alpha \leq \sup_{x \geq 0} \mid \bar{F}(x) - e^{-\lambda x} \mid \leq C_2\epsilon^{w_k},$$

where

$$w_k = \begin{cases} 1 - \delta_k, & \text{for } 1 \leq k \leq 4, \\ (1 - \delta_k) \prod_{m=1}^{\lfloor \frac{k}{4} \rfloor} \left(1 - \frac{\cos(2\pi m/k)}{\cos(2\pi(m-1)/k)}\right), & \text{for } k \geq 5, \end{cases}$$

and $\delta_k \to 0$ as $\epsilon \to 0$.

Let us again cite a series of results related to the characterization of the exponential distribution by an independent function of record values that were introduced in Section 3.

Theorem 4.6. Let $\{\xi_i\} \in P$ be independent, identically distributed random variables with common distribution function $F(x)$ which is absolutely continuous. Then a necessary and sufficient condition for $\xi_i \in \mathcal{E}$ is the independence of the random variables $\xi_{L(n-1)}$ and ς_n.

Proof. It follows from formula (3.8) that if $\xi \in \mathcal{E}, i = 1, 2, ...$, the joint distribution of ς_n and $\xi_{L(n-1)}$ has density

$$f(z, u) = \begin{cases} u^{n-1} \lambda^{n+1} \exp\{-\lambda(z + u)\}/\Gamma(n), & 0 < u,\ z < \infty, \\ 0, & \text{otherwise.} \end{cases}$$

Hence it follows immediately that ς_n and $\xi_{L(n-1)}$ are independent.

Let us now assume that ς_n and $\xi_{L(n-1)}$ are independent. The density of the joint distribution of ς_n and $\xi_{L(n-1)}$ is given by formula (3.8) and the density of the distribution of $\xi_{n(n-1)}$ is

$$f(u) = \begin{cases} R^{n-1}(u) f(u)/\Gamma(n), & 0 < u < \infty, \\ 0, & \text{otherwise.} \end{cases} \tag{4.13}$$

From the independence of ς_n and $\xi_{L(n-1)}$ and formulas (3.8) and (4.13) we get

$$f(u + z)/F(u) = g(z)$$

where $g(z)$ is the density of ς_n. Integrating the identity with respect to z from 0 to z_1 we obtain

$$\frac{\bar{F}(u) - \bar{F}(u + z_1)}{\bar{F}(u)} = G(z_1).$$

From this functional equation and Theorem 2.1, we have $\xi_1 \in \mathcal{E}$.

The class $\mathcal{E}$ can be characterized in the same way, using the regression of $\xi_{L(1)}$ on ξ_1.

It is well-known that for $n \geq 1$

$$P\{\xi_{L(n)} \leq x \mid \xi_{L(n-1)} = y\} = [F(x) - F(y)]/[1 - F(y)], \quad x > y.$$

Let $[a, b]$ be the support of $F(\cdot)$, where a and b may be infinite. Let $h(\cdot)$ be a strictly increasing function mapping $[a, b]$ onto $[c, d]$ (and therefore continuous on $[a, b]$), where $-\infty \leq c < d \leq \infty$. Let us assume that $\int_y^b h(x)dF(x)$ exists for $y > a$, and note that the regression function $E(h(\xi_{L(1)}) \mid \xi_{L(0)} = y)$ is defined with probability one with respect to the distribution of the random variable $\xi_{L(0)}$.

Theorem 4.7. If

$$E\{h(\xi_{L(1)}) \mid \xi_{L(0)} = y\} = K(y) \tag{4.14}$$

with probability one with respect to $F(x)$, where $K(y)$ is a non- decreasing function on $[c, d]$, then $F(x)$ is uniquely defined.

Proof. It is easy to show that $E\{h(\xi_{L(1)}) \mid \xi_{L(0)} = y\} = K(y)$ is equivalent to $E\{\eta_{L(1)} \mid \eta_{L(0)} = u\} = L(u)$ where $L(u)$ is a non-decreasing function on $[c, a]$ with $\eta = h(\xi)$. Therefore, we can consider only the case $h(\xi) = \xi$ and prove that if

$$E\{\xi_{L(1)} \mid \xi_{L(0)} = y\} = K(y)$$

and $K(y)$ is a non-decreasing function on $[a, b]$, then $F(\cdot)$ is uniquely defined.

In as much as $E\{\xi_{L(1)} \mid \xi_{L(0)} = y\} = K(y)$ we can choose

$$R(y) = \begin{cases} [1 - F(y)]^{-1} \int_y^b x\,dF(x), & y < b, \\ y, & y \geq b. \end{cases} \tag{4.15}$$

39

Then $R(\cdot)$ is a non-decreasing, continuous function, $F(y_1) < F(y_2)$ implies $R(y_1) < R(y_2)$ and $F(\cdot)$ is uniquely defined since in fact

$$1 - F(y) = S(a)[S(y)]^{-1}\exp\{-\int_a^y [S(x)]^{-1}dx\}, \quad y \in [a, b],$$

where $S(x) = R(x) - x$.

From (4.14) and (4.15) we have

$$R(y) = K(y) \tag{4.16}$$

with probability one with respect to $F(\cdot)$. If there exists a subinterval $[a, b]$ on which $F(\cdot)$ is constant, then it follows from (4.16) that $R(y) = K(y)$ for all $y \in [a, b]$. If $F(\cdot)$ is constant on $(\ell, m) \subset [a, b]$, where ℓ and m are points of increase of $F(\cdot)$, then $R(\cdot)$ is constant on $[\ell, m]$. Using this fact and (4.16) we have $K(\ell) = K(m) = R(\ell)$. Since $K(\cdot)$ is non-decreasing, $R(y) = K(y)$ for all $y \in [\ell, m]$. This implies $R(y) = K(y)$ for all $y \in [a, b]$. Hence it follows that $K(\cdot)$ defines $R(\cdot)$ uniquely, and therefore $F(\cdot)$. ||

Corollary 4.2. If $E\{\xi_{L(1)} \mid \xi_{L(0)} = y\} = y$, then $\xi \in \mathcal{E}$.

Let us formulate yet another characterizing property of the exponential distribution.

Theorem 4.8. If $\xi \in \mathcal{P}$ and

$$k = D\{(\xi - y) \mid \xi \geq y\} = D\xi = \text{const.}, \quad 0 \leq y < \infty,$$

then $\xi \in \mathcal{E}$.

Proof. It is easy to show that

$$\begin{aligned}
k = D\xi &= D\{(\xi - y) \mid \xi \geq y\} \\
&= \frac{2}{\bar{F}(y)}\int_y^\infty (x - y)\bar{F}(x)dx - [\frac{1}{\bar{F}(y)}\int_y^\infty \bar{F}(x)dx]^2.
\end{aligned} \tag{4.17}$$

40

Letting $\Psi(y) = \int_y^\infty (x - y)\bar{F}(x)dx$ and taking into account that

$$\Psi'(y) = -\int_y^\infty \bar{F}(x)dx, \quad \Psi''(y) = \bar{F}(y),$$

we obtain from (4.17),

$$k = \frac{2\Psi(y)}{\Psi''(y)} - [\frac{\Psi'(y)}{\Psi''(y)}]^2.$$

Setting $P(\Psi) = \Psi'(y)$, we have

$$P(P')^2 - \frac{2}{k}\Psi P' + \frac{1}{k}P = 0. \qquad (4.18)$$

Substituting $P^2(\Psi) = z(\Psi)$, (4.18) becomes Clairault's equation:

$$(z')^2 - \frac{4}{k}\Psi z' + \frac{4}{k}z = 0.$$

Solutions of this differential equation have the form

$$\Psi_1(y) = c_0 e^{\pm y\sqrt{\frac{1}{k}}}, \quad \Psi_2(y) = -\frac{4}{k}(2y + c_1)^2 - c_2,$$

where c_0, c_1, c_2 are constants. Only $\Psi_1(y)$ meets our requirements; therefore

$$\Psi(y) = \Psi(0)e^{-y\sqrt{\frac{1}{k}}},$$

where $\Psi(0) = c_0$.

Correspondingly, for $\bar{F}(y)$ we obtain

$$\bar{F}^2(y) - \frac{2}{k}\Psi(y)\bar{F}(y) + \frac{1}{k}[\Psi'(y)]^2 = 0,$$

which has the solution

$$\bar{F}_{1,2}(y) = \frac{1}{k}\Psi(y) = \frac{\Psi(0)}{k}e^{-y\sqrt{\frac{1}{k}}}.$$

From the condition $\bar{F}(0) = 1$, we find that $c_0 = \Psi(0) = k$. Thus, $\bar{F}(y) = e^{-y\sqrt{\frac{1}{k}}}$. ||

41

Chapter 5

CHARACTERIZATION OF THE EXPONENTIAL DISTRIBUTION BY A PROPERTY OF ITS ORDER STATISTICS

Let $\{\xi_i\}$ be independent, identically distributed random variables with distribution function $F(x)$. Assume that $m_1, m_2, \ldots, m_n (n \geq 1)$ are positive integers and $c, a_1, a_2, \ldots, a_n$ are positive numbers such that $c > \max\{a_k\}$ if $n > 1$ and $c = a_1$ if $n = 1$. Let α be a unique positive number that satisfies $a_1^\alpha + a_2^\alpha + \cdots + a_n^\alpha = c^\infty$ if $n > 1$ and an arbitrary positive number if $n = 1$.

Let us consider $\xi_{1,m_k}, k = 1, \ldots, n$ and define

$$\varsigma = \min_{1 \leq k \leq n} \{(cm_k^{1/\alpha}/a_k)\xi_{1,m_k}\}.$$

We shall not consider the case $n = a_1 = m_1 = 1$.

Theorem 5.1. ς has distribution function $F(x)$ if and only if there exists a positive, bounded periodic function $H(x)$ with period $A_k = \log cm_k^{1/\alpha}/a_k$, $k = 1, \ldots, n$ and $F(x)$ can be represented in the form

$$F(x) = \begin{cases} 0, & x < 0, \\ 1 - \exp\{-H(-\log x)x^\alpha\}, & x \geq 0. \end{cases} \tag{5.1}$$

Proof. Using the monotone transformation $\xi_i \longrightarrow (c\xi_i)^\alpha$, the problem is reduced to the case $\alpha = c = 1$, which we shall do, without loss of generality. Then

$$a_1 + a_2 + \cdots + a_n = 1. \tag{5.2}$$

From the defintion of ς we have, for any $x > 0$

$$P\{\varsigma \geq x\} = P\{\frac{m_k}{a_k}\xi_{1,m_k} \geq x, \quad k = 1, 2, \ldots, n\}$$
$$= \prod_{k=1}^{n} P\{\xi_{1,m_k} \geq a_k x/m_k\} = \prod_{k=1}^{n} \bar{F}^{m_k}(a_k x/m_k). \tag{5.3}$$

If $F(x)$ can have form (5.1), then (5.3) becomes

$$P\{\varsigma \geq x\} = \prod_{k=1}^{n} \exp\{-m_k H(-\log\frac{a_k x}{m_k})\frac{a_k x}{m_k}\}$$
$$= \exp\{-H(-\log x)x\} = \bar{F}(x),$$

which we needed to prove.

Assume now that ς has distribution function $F(x)$. Then, from (5.3) it follows that

$$\bar{F}(x) = \prod_{k=1}^{n} \bar{F}^{m_k}(\frac{a_k x}{m_k}), \quad x \geq 0. \tag{5.4}$$

Since $a_k/m_k < 1$ for $k = 1, \ldots, n$, it follows from (5.4) that $F(x) < 1$ for all $x > 0$. Then the function

$$H(x) = -e^x \log(1 - F(e^{-x}))$$

is defined for all real x and satisfies the functional equation

$$H(x) = \sum_{k=1}^{n} a_k H(x + A_k), \quad -\infty < x < \infty. \tag{5.5}$$

Let x_0 be any real number. From (5.5) we have

$$c \equiv \sup_{x \geq x_o} H(x) < \infty \tag{5.6}$$

and

$$H(x + A_k) = H(x), \quad k = 1, \ldots, n, \quad x \geq x_0. \tag{5.7}$$

43

This will follow from Theorem II.1. If (5.6) is established, then, applying Choquet-Deny's Theorem (see [24], v. 2), we get (5.7). However, since (5.5) is a special case of the equation studied in Theorem II.1, the proof can be simplified. To prove (5.6) it is sufficient to show that the inequality

$$H(x) \le e^{A} H(x_0) \tag{5.8}$$

holds for all $x \ge x_0$, where $A = \max\{A_1, A_2, \ldots, A_n\}$. The existence of a k_1 such that $H(x_0) \ge H(x_0 + A_{k_1})$ follows from (5.5). Analogously, there exists a k_2 such that $H(x_0) \ge H(x_0 + A_{k_1}) \ge H(x_0 + A_{k_1} + A_{k_2})$. Proceeding in this way we shall obtain a sequence $k_1, k_2, \ldots$ of positive integers such that $H(x_0) \ge H(x_0 + A_{k_1} + \cdots + A_{k_m})$, $m = 1, 2, \ldots$. If $x_0 \le x \le x_0 + A$, then (5.8) follows from

$$H(x + y) \le e^{y} H(x), \quad x \ge x_0, \quad y \ge 0, \tag{5.9}$$

which can be proved using the definition of $H(x)$.

Let $x > x_0 + A$. Since $\min\{A_1, \ldots, A_n\} > 0$, we can find an m such that

$$x_0 + A_{k_1} + \cdots + A_{k_m} \le x < x_0 + A_{k_1} + \cdots + A_{k_{m+1}}.$$

Setting $\delta \equiv x - (x_0 + A_{k_1} + \cdots + A_{k_m})$, from (5.9) we obtain

$$H(x) = H(x_0 + A_{k_1} + \cdots + A_{k_m} + \delta)$$
$$\le e^{\delta} H(x_0 + A_{k_1} + \cdots + A_{k_m}) \le e^{A} H(x_0),$$

which we were required to prove.

In order to complete the proof of the theorem, it is sufficient to prove (5.7) only for $k = 1$. For this let us introduce a bounded function $K(x) \equiv H(x + A_1) - H(x)$. It is easy to show that $K(x)$ satisfies the functional equation:

$$K(x) = \sum_{k=1}^{n} a_k K(x + A_k), \quad x \ge x_0. \tag{5.10}$$

44

Iterating (5.10) we obtain

$$K(x) = \sum_{k_1 + \cdots + k_n = m} \frac{m!}{k_1! \cdots k_n!} a_1^{k_1} \cdots a_n^{k_n} K(x + k_1 A_1 + \cdots + k_n A_n), \qquad (5.11)$$

where m is any positive integer and the summation is carried out over all $k_i \geq 0, i = 1, \ldots, n, k_1 + \cdots + k_n = m$.

Note that each of the $k_1 A_1 + \cdots + k_n A_n$ on the right-hand side of (5.11) is not less than $m \min\{A_1, \ldots, A_n\}$ which tends to ∞ as $n \longrightarrow \infty$. In particular, for any $\epsilon > 0$ one can choose a sufficiently large m such that

$$\begin{aligned}
| K(x) | &= \sum \frac{m!}{k_1! \cdots k_n!} a_1^{k_1} a_2^{k_2} \cdots a_n^{k_n} | K(x + k_1 A_1 + \cdots + k_n A_n) | \\
&\leq \sum \frac{m!}{k_1! \cdots k_n!} a_1^{k_1} a_2^{k_2} \cdots a_n^{k_n} (\epsilon + \overline{\lim_{x \to \infty}} | K(x) |) \\
&\leq (a_1 + \cdots + a_n)^m (\epsilon + a) = \epsilon + a,
\end{aligned}$$

where $a = \overline{\lim_{x \to \infty}} | K(x) |$. Since $\epsilon > 0$ is arbitrary, we have that

$$| K(x) | \leq a, \quad x \geq x_0. \qquad (5.12)$$

Let us prove (5.7), showing that $a = 0$. Without loss of generality set $a = \overline{\lim_{x \to \infty}} K(x) \geq - \lim_{x \to \infty} K(x)$. From (5.10) and (5.12) it follows that

$$K(x) \leq a_1 K(x + A_1) + (1 - a_1)a, \quad x \geq x_0.$$

Iterating this inequality we have

$$K(x) \leq a_1^k K(x + k A_1) + (1 - a_1^k)a, \quad x \geq x_0, \quad k = 0, 1, 2, \ldots . \qquad (5.13)$$

From the definition of a it follows that there exists an $x_1 (> x_0)$ such that for any $\epsilon > 0$ and any positive integer L, $a - \epsilon a_1^L \leq K(x_1)$ holds. From (5.13) we have

$$a - \epsilon a_1^L \leq K(x_1) \leq a_1^k K(x_1 + k A_1) + (1 - a_1^k)a, \quad k = 0, 1, 2, \ldots,$$

45

which implies

$$a - \epsilon a_1^{L-k} \leq K(x_1 + kA_1), \quad k = 0, 1, \dots .$$ (5.14)

Adding both sides of (5.14) for $k = 0, 1, \dots, L-1$ we get

$$L(a - \epsilon) \leq \sum_{k=0}^{L-1}(a - \epsilon a_1^{L-k}) \leq \sum_{k=0}^{L-1} K(x_1 + kA_1)$$
$$= H(x + LA_1) - H(x_1) \leq 2 \sup \mid H(x) \mid = 2c.$$

For arbitrary ϵ and L this is possible only if $a = 0$. The reasoning does not depend on the choice of x_0 and the proof of the theorem is completed.$\|$

Corollary 5.1. Let $\xi_1, \dots, \xi_n$ be independent, identically distributed random variables with common distribution function $F(x)$ and let $a_1, a_2, \dots, a_n$ be positive constants satisfying condition (5.2) and such that $\log a_i / \log a_j$ is irrational for some i and j. If $\varsigma = \min_{1 \leq k \leq n} \{\xi_k / a_k\}$ has the same distribution function $F(x)$ as ξ_1, then $\xi_1 \in \mathcal{E}$.

Corollary 5.2. Assume that the $\xi_i \in \mathcal{P}$ are independent and identically distributed. If $n_1 \xi_{1,n_1}$ has the same distribution as $n_2 \xi_{1,n_2}$ for n_1 and n_2 such that $\log n_1 / \log n_2$ is irrational, then $\xi_1 \in \mathcal{E}$.

Proof of the Corollaries. To these two cases there correspond an $m_i = 1, i = 1, \dots, n$, and an $n = a_1 = 1 < m_j, j = 1, 2$. It follows from Theorem 5.1 that $F(x)$ has the form (5.1) and $H(x)$ has period $A_j = -\log a_j, j = 1, 2, \dots, n$ (in Corollary 5.1) and $A_j \equiv \log m_j, j = 1, 2$, (in Corollary 5.2). Since A_i / A_j is irrational in both cases, $H(x)$ must be a constant. $\|$

Corollary 5.3. Let the function $F(x)$ be such that $\lim_{x \to 0} F(x)/x = \lambda > 0$. If $m\xi_{1,m}$ has distribution $F(x)$ for some $m > 1$, then $F(x)$ is exponential.

Proof. From Theorem 5.1 $F(x)$ can have form (5.1). Since $\lim_{x \to \infty} H(x) = \lambda$ and $H(x)$ is periodic, $H(x)$ must be a constant. $\|$

Let $\xi_1, \xi_2, \ldots, \xi_n$ be a sample of size n from a general population with distribution function $F(x)$ and density $f(x)$ and let $\xi_{1,n} \leq \xi_{2,n} \leq \cdots \leq \xi_{n,n}$ be the order statistics associated with this sample. In Section 3 the variables

$$D_{r,n} = (n - r)(\xi_{r+1,n} - \xi_{r,n}), \ 1 \leq r < n$$

with

$$D_{0,n} = n\xi_{1,n}, \quad D_{n,n} = 0$$

were introduced and it was shown that if the $\{\xi_i\}_{i=1}^n$ are independent, identically and exponentially distributed random variables, then $D_{r,n} \sim \xi_1$ ($\sim$ denotes the same distribution). It turns out that this property, with certain restrictions on the distribution of ξ_1, characterizes the exponential distribution.

If the failure rate density $r(x) = f(x)/[1 - F(x)]$ is either a decreasing or an increasing function, then we shall say that $F(x)$ belongs to class C.

Theorem 5.2. Assume that $\xi \in P$ has an absolutely continuous distribution function $F(x)$ which is strictly increasing on $[0, \infty)$. Then the following properties are equivalent:

(a) $\xi \in \mathcal{E}$;

(b) for some i and n, $1 \leq i < n$, the statistic $D_{i,n} \sim \xi_1$ and $F(x) \in C$.

Proof. In Section 3 it was shown that (a) $\Longrightarrow$ (b). Let us prove that (b) $\Longrightarrow$ (a).

It is well-known that the density of the distribution of $D_{i,n}$ has the form

$$f_{D_{i,n}}(z) = \frac{n!}{(i-1)!(n-i-1)!} \int_0^\infty F^{i-1}(u)[1 - F(u + \frac{z}{n-i})]^{n-i-1} f(u) f(u + \frac{z}{n-i}) \frac{du}{n-i}.$$
$$(5.15)$$

By assumption $f_{D_{i,n}}(z) = f(z)$, so writing

$$\frac{(i-1)!(n-i)!}{n!} = \int_0^\infty F^{i-1}(u)(1 - F(u))^{n-i} f(u) du,$$

47

we get

$$0 = \int_0^\infty F^{i-1}(u)f(u)g(u,z)du, \quad \text{for all } z, \tag{5.16}$$

where

$$g(u,z) = f(z)(1 - F(u))^{n-i} - (1 - F(u + \frac{z}{n-i}))^{n-i-1}f(u + \frac{z}{n-i}).$$

Integrating (5.16) with respect to z from 0 to z_1 and changing the order of integration we obtain

$$0 = \int_0^\infty F^{i-1}(u)(1 - F(u))^{n-i}f(u)G(u,z_1)du, \quad \text{for all } z_1, \tag{5.17}$$

where

$$G(u,z) = [\bar{F}(u + \frac{z_1}{n-i})/\bar{F}(u)]^{n-i} - \bar{F}(z_1).$$

Furthermore, if $r(x)$ is an increasing function, then for any integer $k > 0$, $\bar{F}(x/k) \geq \bar{F}^{1/k}(x)$, so that $G(0, z_1) \geq 0$. Thus, if (5.17) is satisfied, then it is necessary that $G(0, z_1) \equiv 0$. The reasoning is analogous for the case when $r(x)$ is a non-increasing function. Rewriting $G(0, z_1)$ in terms of $F(z_1)$ we have

$$1 - F(z_1) = [1 - F(\frac{z_1}{n-i})]^{n-i}, \quad \text{for all } z_1. \tag{5.18}$$

Setting $n - i = k$, in (5.18), we obtain

$$\bar{F}(\frac{z_1}{k}) = \bar{F}^{1/k}(z_1), \quad \text{for all } z_1 > 0 \text{ and some integer } k > 0. \tag{5.19}$$

The solution of (5.19) for $k > 1$ is (see [26])

$$\bar{F}(z_1) = 1 - F(z_1) = e^{-\lambda_1 z_1}, \quad \text{for some } \lambda_1 > 0 \text{ and for all } z_1 > 0. \tag{5.20}$$

If $k = 1$, then $G(u, z_1) = [\bar{F}(u + z_1)/\bar{F}(u)] - \bar{F}(z_1)$ and (5.17) gives

$$0 = \int_0^\infty F^{n-2}(u)f(u)[\frac{\bar{F}(u + z_1)}{\bar{F}(u)} - \bar{F}(z_1)]\bar{F}(u)du \quad \text{for all } z_1 \text{ and } F(x) \in C.$$

48

This means that $\bar{F}(u + z_1) = \bar{F}(u)\bar{F}(z_1)$, which again gives (5.20).‖

Theorem 5.3. Assume that $\xi \in P$ has an absolutely continuous distribution function $F(x)$ which is strictly increasing on $[0, \infty)$. Then the following properties are equivalent:

(a) $\xi \in \mathcal{E}$;

(b) for some i, j and n, $1 \le i < j < n$, and $D_{i,n} \sim D_{j,n}$, $F(x)$ belongs to the class C.

Proof. The assertion (a) $\Longrightarrow$ (b) was proven in Section 3. Let us prove that (b) $\Longrightarrow$ (a). From the representation of the joint conditional density of $\xi_{j,n}$ and $\xi_{j+1,n}$, given that $\xi_{i,n} = x$, it follows that the conditional density of $D_{j,n}$, given $\xi_{i,n} = x$ is

$$f_{D_{j,n}}(d \mid \xi_{i,n} = x)$$
$$= K \int_0^\infty \left[\frac{\bar{F}(x) - \bar{F}(x+s)}{\bar{F}(x)} \right]^{j-i-1} \left(\frac{\bar{F}(x+s+\frac{d}{n-j})}{\bar{F}(x)} \right)^{n-j-1} \frac{f(x+s)}{\bar{F}(x)} \frac{f(x+s+\frac{d}{n-j})}{\bar{F}(x)} ds, \tag{5.21}$$

where $K = (n-i)!/[(j-i-1)!(n-j)!]$ and $1 \le i < j < n$.

Integrating (5.21) with respect to d from d to ∞ we get

$$F_{D_{j,n}}(d \mid \xi_{i,n} = x)$$
$$= K \int_0^\infty \left[\frac{\bar{F}(x) - \bar{F}(x+s)}{\bar{F}(x)} \right]^{j-i-1} \frac{f(x+s)}{\bar{F}(x)} \left[\frac{\bar{F}(x+s+\frac{d}{n-j})}{\bar{F}(x)} \right]^{n-j} ds. \tag{5.22}$$

Furthermore, we know that the conditional distribution of $D_{i,n}$ satisfies the relation

$$\bar{F}_{D_{i,n}}(d \mid \xi_{i,n} = x) = \left[\frac{\bar{F}(x + \frac{d}{n-i})}{\bar{F}(x)} \right]^{n-i}, \quad 1 \le i < n. \tag{5.23}$$

Taking into account that

$$K^{-1} = \int_0^\infty \left[\frac{\bar{F}(x+s)}{\bar{F}(x)} \right]^{n-j} \left[\frac{\bar{F}(x) - \bar{F}(x+s)}{\bar{F}(x)} \right]^{j-i-1} \frac{f(x+s)}{\bar{F}(x)} ds,$$

49

and simplifying (5.22) and (5.23), we get

$$0 = \int_0^\infty \left[\frac{\bar{F}(x+s)}{\bar{F}(x)} \right]^{n-j} \left[\frac{\bar{F}(x) - \bar{F}(x+s)}{\bar{F}(x)} \right]^{j-i-1} \frac{f(x+s)}{\bar{F}(x+s)} G(x,s,d) ds \qquad (5.24)$$

for all d and for any given x where

$$G(x,s,d) = \left[\frac{\bar{F}(x + \frac{d}{n-i})}{\bar{F}(x)} \right]^{n-i} - \left[\frac{\bar{F}(x + s + \frac{d}{n-j})}{\bar{F}(x+s)} \right]^{n-j}.$$

Differentiating $G(x,s,d)$ with respect to s we have

$$\frac{\partial}{\partial s} G(x,s,d) = \left[\frac{\bar{F}(x + s + \frac{d}{n-j})}{\bar{F}(x+s)} \right]^{n-j} [r(x + s + \frac{d}{n-j}) - r(x+s)].$$

i) If $r(x)$ is an increasing function then $G(x,s,d)$ increases with s for fixed x and d. Thus, in order that (5.24) be satisfied, we must have $G(x,0,d) \leq G(x,s,d) \leq 0$. If $r(x)$ does not decrease then we know (see property 2.1) that $\log \bar{F}(x)$ is a convex function, and using Jensen's inequality we get

$$\log \bar{F}(x + \frac{d}{n-i}) \geq \frac{j-i}{n-i} \log \bar{F}(x) + \frac{n-j}{n-i} \log \bar{F}(x - \frac{d}{n-j}),$$

that is

$$\bar{F}[(x + \frac{d}{n-i})]^{n-1} \geq \bar{F}^{j-i}(x)\bar{F}^{n-j}(x + \frac{d}{n-j}).$$

This inequality shows that $G(x,0,d) \geq 0$. Therefore, if (5.24) is correct, then we must have $G(x,0,d) = 0$ for all d and for a given x.

(ii) If $r(x)$ is a non-increasing function then, proceeding completely analogously, we get $G(x,0,d) = 0$ for all d and for any given x. Substituting $x = 0$, we will have $G(0,0,d) = 0$ for all d, that is,

$$[\bar{F}(\frac{d}{n-i})]^{n-i} = [\bar{F}(\frac{d}{n-j})]^{n-j}$$

for all $d \geq 0$ and for some i, j, and n $(1 \leq i < j < n)$. Setting $\varphi(d) = -\log \bar{F}(d)$ and $z = d/(n-i)$ we get

$$\varphi(z) = \frac{n-j}{n-i} \; \varphi(\frac{n-i}{n-j} z) \tag{5.25}$$

for all $z \geq 0$ and some i, j and n with $1 \leq i < j < n$.

The non-trivial solution of (5.25) is $\varphi(z) = cz$, where c is a constant. Therefore, $F(x) = 1 - e^{-cx}$. Using the boundary conditions $F(0) = 0$ and $F(\infty) = 1$ we have

$$F(x) = 1 - e^{-\theta x} \quad \text{where} \quad \theta > 0. \; \|$$

Let us denote $V_i = \xi_{s_i,n} - \xi_{r,n}$.

Theorem 5.4. Let $\xi \in \mathcal{P}$ have an absolutely continuous, strictly increasing distribution function $F(x)$. The following assertions are equivalent:

(a) $\xi \in \mathcal{E}$;

(b) $V_i \sim \xi_{s_i-r,n-r}$ for $i = 1, 2$ and for any fixed r; s_1 and s_2 are distinct numbers $(1 < r < s_1 < s_2 \leq n)$.

Proof. (a) $\Longrightarrow$ (b) was proved in Section 3. Let us prove (b) $\Longrightarrow$ (a).

The density of the joint distribution of $\xi_{r,n}$ and V_i has the form

$$f_{\xi_{r,n}V_i}(u, v_i) = K_i F^{r-1}(u)[F(u+v_i) - F(u)][1 - F(u+v_i)]^{n-s_i} f(u) f(u+v_i),$$

for $0 \leq u < \infty$, $0 \leq v_i < \infty$, where

$$k_i = \frac{n!}{(r-1)!(s_i-r-1)!(n-s_i)!}.$$

The density of V_i is obtained from (5.26) by integrating with respect to u, so we get

$$f_{V_i}(v_i) = \int_0^\infty f_{\xi_{r,n},V_i}(u, v_i) du, \tag{5.27}$$

51

and the density of $\xi_{s_i-r,n-r}$ is

$$f_{\xi_{s_i-r,n-r}}(\omega) = (s_i - r)\binom{s_i - r}{n - r} F^{s_i-r-1}(\omega)(1 - F(\omega))^{n-s_i} f(\omega), \qquad (5.28)$$

for $0 \le \omega < \infty$. Since V_i and $\xi_{s_i-r,n-r}$ have the same distribution, from (5.27) and (5.28) we get the integro-differential equation

$$\begin{aligned}
F^{s_i-r-1}&(x)[1 - F(x)]^{n-s_i} f(x) \\
&= r\binom{n}{r} \int_0^\infty F^{r-1}(u)[F(u + x) - F(u)]^{s_i-r-1}[1 - F(u+x)]^{n-s_i} f(u)f(x + u)du.
\end{aligned} \qquad (5.29)$$

Since $F(x)$ and $1 - F(x)$ are strictly positive for all x, then, dividing (5.29) by

$$[F(x)]^{s_i-r-1}[1 - F(x)]^{n-s_i} f(x)$$

we get

$$r\binom{n}{r} = \int_0^\infty F^{r-1}(u)\left[\frac{F(u + x) - F(u)}{F(x)}\right]^{s_i-r-1}\left[\frac{1 - F(u + x)}{1 - F(x)}\right]^{n-s_i}\frac{f(u)}{f(x)}f(u + x)du \quad (5.30)$$

for $i = 1, 2$ and all $x > 0$.

Let us rewrite (5.30) in the form

$$\begin{aligned}
\int_0^\infty F^{r-1}(u)&\left[\frac{F(u + x) - F(x)}{1 - F(u + x)}\,\frac{1 - F(x)}{F(x)}\right]^{s_i}\left[\frac{1 - F(u + x)}{1 - F(x)}\right]^{n} \\
&\times \left(\frac{F(x)}{F(u + x) - F(u)}\right)^{r+1}\left(\frac{f(u)f(u + x)}{f(x)}\right)du = r\binom{n}{r}.
\end{aligned} \qquad (5.31)$$

The right-hand side of (5.31) does not depend on s_i and so the integral on the left-hand side does not depend on $s_i, i = 1, 2$, either. Set

$$G \equiv G(u, x) = \frac{F(u + x - F(u)}{1 - F(u + x)}\,\frac{1 - F(x)}{F(x)}$$

52

and take $i = 1$ and $i = 2$ in (5.31). Then, subtracting the expression for $i = 2$ from the expression obtained for $i = 1$, we get

$$0 = \int_0^\infty F^{r-1}(u)[G^{s_1} - G^{s_2}] \left[\frac{1 - F(u + x)}{1 - F(x)}\right]^n \left[\frac{F(x)}{F(u + x) - F(u)}\right]^{r+1} \frac{f(u)f(u + x)}{f(x)} du. \quad (5.32)$$

Since $F(x)$ is monotone increasing for all x, we have from (5.32)

$$G^{s_1} - G^{s_2} = G^{s_1}(1 - G^h) = 0, \quad h = s_2 - s_1.$$

Therefore either $G = 0$ or $G = 1$. Since neither $F(u + x) - F(u)$ nor $1 - F(x)$ can be equal to zero under the conditions of the theorem, $G \neq 0$. Therefore, $G = 1$ and we have

$$[F(u + x) - F(u)][1 - F(x)] = F(x)[1 - F(u + x)]$$

or $\bar{F}(u + x) = \bar{F}(u)\bar{F}(x)$. The solution of this equation is

$$\bar{F}(x) = e^{-\lambda x}, \quad x \geq 0, \quad \lambda > 0. \parallel$$

Let us repeat that the distribution function $F(x)$ belongs to class C_1 if either $\bar{F}(x + y) \leq \bar{F}(x)\bar{F}(y)$ or $\bar{F}(x + y) \geq \bar{F}(x)\bar{F}(y)$ for all $x, y \geq 0$.

In Section 3 a sequence of record values $\{\xi_{L(n)}, n \geq 0\}$ and the random variable ς_n were defined.

Theorem 5.5. Let $\{\xi_n\} \in \mathcal{P}$ be a sequence of independent, identically distributed random variables with distribution function $F(x)$ and density $f(x)$. If $F(x) \in C_1$ and $\varsigma_n \sim \xi_1$, then $\xi_1 \in \mathcal{E}$.

Proof. By virtue of (3.8) the density of ς_n can be written as:

$$f_{\varsigma_n}(z) = \begin{cases} \int_0^\infty R^{n-1}(u) \ r(u)f(u + z)du/\Gamma(n), & 0 < z < \infty, \\ 0, & \text{otherwise.} \end{cases}$$

53

From the assumption that ς_n and ξ_1 have the same distribution we have

$$\int_0^\infty R^{n-1}(u)\ r(u)f(u+z)\frac{du}{\Gamma(n)} = f(z) \quad \text{for all } z > 0.$$

From

$$\int_0^\infty R^{n-1}(u)f(u)du = \Gamma(n)$$

we have

$$\int_0^\infty R^{n-1}(u)\ r(u)f(u+z)du = f(z)\int_0^\infty R^{n-1}(u)f(u)du \quad \text{for all } z > 0,$$

that is

$$\int_0^\infty R^{n-1}(u)f(u)[\frac{f(u+z)}{\bar{F}(u)} - f(z)]du = 0 \quad \text{for all } z > 0. \tag{5.33}$$

Integrating (5.33) with respect to z from z_1 to ∞, we obtain

$$\int_0^\infty R^{n-1}(u)f(u)[\frac{\bar{F}(u+z_1)}{\bar{F}(u)} - \bar{F}(z_1)]du = 0 \quad \text{for all } z_1 > 0.$$

Since $F(x) \in C_1$, the last relation can hold if and only if

$$\bar{F}(u+z_1)/\bar{F}(u) = \bar{F}(z_1) \quad \text{for all } z_1 > 0.$$

From this it follows that $\xi_1 \in \mathcal{E}$. $\|$.

Theorem 5.6. Let $\{\xi_k\} \in \mathcal{P}$ be a sequence of independent, identically distributed random variables with distribution function $F(x)$ and density $f(x)$. If $F(x) \in C$ and $\varsigma_n \sim \varsigma_{n+1}$, then $\xi_1 \in \mathcal{E}$.

Proof. From (3.8) it follows that

$$P\{\varsigma_n \geq x\} = \begin{cases} \int_0^\infty R^{n-1}(u)\ r(u)\bar{F}(u+z)du/\Gamma(n), & \text{for } z \geq 0, \\ 0, & \text{otherwise.} \end{cases}$$

54

Since ς_n and ς_{n+1} are identically distributed,

$$\int_0^\infty R^n(u)\ r(u)\bar{F}(u+z)du = n\int_0^\infty R^{n-1}(u)\ r(u)\bar{F}(u+z)du \text{ for } z \geq 0. \tag{5.34}$$

But

$$n\int_0^\infty R^{n-1}(u)\ r(u)\bar{F}(u+z)du = \int_0^\infty R^n(u)f(u+z)du. \tag{5.35}$$

Substituting (5.35) in (5.34) and simplifying, we get

$$\int_0^\infty R^{n-1}(u)\ r(u)\bar{F}(u+z)[1 - \frac{r(u+z)}{r(u)}]du = 0 \text{ for } z \geq 0.$$

Since $F(x) \in C$ the expression obtained is true if and only if

$$r(u+z) = r(u) \quad \text{for all} \quad u, z \geq 0.$$

Therefore, $\xi_1 \in \mathcal{E}$. $\|$

Chapter 6

CHARACTERIZATIONS OF DISTRIBUTIONS BY MOMENT PROPERTIES OF ORDER STATISTICS

Let $\xi_{1,n} \leq \xi_{2,n} \leq \cdots \leq \xi_{n,n}$ be the order statistics from n independent observations of the random variable ξ with distribution function $F(x)$.

Let us study the conditions under which the set of moments of the order statistics uniquely defines the distribution function $F(x)$, and look at the exponential distribution as an example. The notation and definitions of Section 3 will be used.

Let $C[0,1]$ be the space of all continuous functions $f(x)$ on the interval $[0,1]$. S. N. Bernstein's theorem asserts that for every $f \in C[0,1]$, the sequence of polynomials

$$B_n(f,x) = \sum_{k=0}^{n} f(\frac{k}{n}) \binom{n}{k} x^k (1-x)^{n-k}$$

converges uniformly to $f(x)$ on $[0,1]$. A sequence of polynomials $\{P_n(x)\}_{n=1}^{\infty}$ is called complete in $L[0,1]$ if for arbitrary $f \in L[0,1]$ the identities

$$\int_0^1 f(x) P_n(x) dx = 0, \quad n = 1, 2 \dots$$

imply $f(x) = 0$ almost everywhere in $[0,1]$.

Theorem 6.1. Let $\{n_j\}, j = 1, 2, \dots$ be a sequence of natural numbers tending to infinity and let L be the space generated by the sequence of polynomials

$$x^k (1-x)^{n_j-k}, \quad k = 0, 1, \dots, n_j; \quad j = 1, 2, \dots \ . \tag{6.1}$$

Then the closure $\bar{L} = C[0,1]$ and therefore the sequence of polynomials (6.1) are complete in

56

$L[0, 1]$.

Proof. The space L contains all polynomials $x^k(1-x)^i, k, i = 0, 1, \ldots, n_j$ and, in particular, if the representation in (3.11) is used, L will also contain all the $x^k, k = 0, 1, \ldots, n_j$. Since $n_j \longrightarrow \infty$ as $j \longrightarrow \infty$, it follows from Weirstrauss' theorem that $\bar{L} = C[0, 1]$.

Let us now assume that $f \in L[0, 1]$ also satisfies

$$\int_0^1 f(x) x^k (1-x)^{n_j - k} dx = 0, \quad k = 0, 1, \ldots, n_j; \quad j = 1, 2, \ldots . \tag{6.2}$$

We set

$$F(x) = \int_0^x f(t) dt$$

and introduce the bounded linear functional

$$\varphi_F(g) = \int_0^1 g(x) dF(x) \quad \text{for} \quad g \in L[0, 1].$$

Turning to (6.2) we find that

$$\varphi_F(x^k(1-x)^{n_j - k}) = 0, \quad k = 0, 1, \ldots, n_j; \quad j = 1, 2, \ldots .$$

It follows from the Hahn-Banach theorem that $\varphi_F \equiv 0$ on $C[0, 1]$, so that $f = 0$ almost everywhere on $[0, 1]$. ∥

Theorem 6.2. Let ξ and η be two random variables with $E \mid \xi \mid < \infty$ and $E \mid \eta \mid < \infty$, and let $F(x)$ and $G(x)$ be their respective distribution functions. $F(x) = G(x)$ if and only if there exists a sequence $\{n_j\}$ of positive integers tending to infinity and

$$E \xi_{k, n_j} = E \eta_{k, n_j}, \quad k = 0, 1, \ldots, n_j. \tag{6.3}$$

Proof. The necessity of (6.3) is obvious. Let us prove the sufficiency. For $m = 1$ formula (3.8) takes the form

$$E\xi_{k,n_j} = k\binom{n_j}{k} \int_0^1 F^{-1}(t)t^{k-1}(1-t)^{n_j-k}dt.$$

The same formula also holds for $E\eta_{k,n_j}$. Since $E \mid \xi \mid < \infty$ and $E \mid \eta \mid < \infty, F^{-1}(t)$ and $G^{-1}(t) \in L[0,1]$.

From formula (6.3) we get

$$\int_0^1 [F^{-1}(t) - G^{-1}(t)]t^{k-1}(1-t)^{n_j-k}dt = 0, \quad k = 1,2,\ldots,n_j, \;\; j = 1,2,\ldots \;.$$

It follows from Theorem 6.1 that $F^{-1}(t) = G^{-1}(t)$ almost everywhere on $[0,1]$, and therefore, from the monotonicity of $F(x)$ and $G(x)$, $F(x) = G(x)$. ‖

The general case $m > 1$ is proved in exactly the same way and follows immediately from Theorem 6.2.

Corollary 6.1. Let m be an odd integer and let ξ and η be two random variables with $E \mid \xi \mid^m < \infty, E \mid \eta \mid^m < \infty$, and with distribution functions $F(x)$ and $G(x)$ respectively. $F(x) = G(x)$ if and only if there exists a sequence of natural numbers $\{n_j\}, n_j \longrightarrow \infty$ as $j \longrightarrow \infty$, and

$$E\xi_{k,n_j}^m = E\eta_{k,n_j}^m, \quad k = 0,1,\ldots,n_j.$$

Proof. Since m is odd, the identity $[F^{-1}(t)]^m = [G^{-1}(t)]^m$ implies $F^{-1}(t) = G^{-1}(t)$, and therefore $F(x) = G(x)$. ‖

In Corollary 6.1 m must be odd. The following example shows that Corollary 6.1 is not true for m even.

Example 6.1. Let $F^{-1}(t)$ and $G^{-1}(t)$ be defined as follows:

$$F^{-1}(t) = a > 0, \quad 0 \le t \le 1,$$

58

$$G^{-1}(t) = \begin{cases} -a, & 0 \le t \le \frac{1}{2}, \\ a, & \frac{1}{2} < t \le 1. \end{cases}$$

Then clearly, $F(x) \ne G(x), E \mid \xi \mid^m < \infty, E \mid \eta \mid^m < \infty$ and

$$E\xi_{k,n}^m = E\eta_{k,n}^m = k\binom{n}{k}\int_0^1 a^m t^{k-1}(1-t)^{n-k}dt,$$

since m is an even number.

The well-known theorem of Müntz-Szász states that a sequence of polynomials $\{x^{n_j}\}, j = 1, 2, \ldots$ with $\sum_{j=1}^{\infty} 1/n_j = \infty$ is complete in $L[0,1]$. We shall use this theorem to obtain sufficient conditions for a unique determination of the distribution function $F(x)$ by a subset of moments of its order statistics.

Theorem 6.3. Assume that $\{p_n(x)\}, n = 1, 2, \ldots$, is a sequence of polynomials (the degree of $p_n(x)$ is n) that is complete in $L[0,1]$. If there exists a sequence of integers $\{n_j\}$ such that

$$\sum_{j=1}^{\infty} \frac{1}{n_j} = \infty$$

and

$$x^{n_j} = a_1(j)p_1(x) + \cdots + a_N(j)p_N(x) + \sum_{i=N+1}^{n_j} a_i(j)p_i(x),$$

where the $a_i(j)$ are integers and uniformly bounded for $i = 1, 2, \ldots, N; j = 1, 2, \ldots$, then the sequence $\{p_n(x)\}, n = N+1, N+2, \ldots$ is also complete in $L[0,1]$.

Proof. Since the series $\sum_{j=1}^{\infty} \frac{1}{n_j}$ is divergent and the integers $a_i(j)$ are bounded, it follows that one can choose a subsequence $\{n_{j_p} = n_p\}$ from n_j such that

$$\sum_{p=1}^{\infty} \frac{1}{n_p} = \infty$$

59

and

$$x^{n_p} = a_1 p_1(x) + \cdots + a_N p_N(x) + \sum_{m=N+1}^{n_p} C_m(p) p_m(x).$$

Let us now look at an $f(x) \in L[0,1]$ which satisfies

$$\int_0^1 f(x) p_m(x) dx = 0, \quad m = N+1, N+2, \ldots.$$

Since

$$\int_0^1 x^{n_p} f(x) dx = a_1 \int_0^1 p_1(x) f(x) dx + \cdots + C_N \int_0^1 p_N(x) f(x) dx = v, \quad p = 1, 2, \ldots$$

the assertion follows directly from the Müntz-Szász theorem if $v = 0$. If $v \neq 0$, we have a contradiction.

Without loss of generality we can assume that $n_1 < n_2 < \cdots$ and set

$$\varphi_F(g) = \int_0^1 g(x) dF(x) \quad \text{for} \quad g \in L[0,1],$$

where $F(x) = \int_0^x t^{n_1} f(t) dt$.

Clearly,

$$\varphi_F(1 - x^{n_p - n_1}) = \int_0^1 (1 - x^{n_p - n_1}) dF(x) = \int_0^1 x^{n_1} f(x) dx - \int_0^1 x^{n_p} f(x) dx = 0.$$

Let M be the space generated by $\{1 - x^{n_p - n_1}\}, p = 2, 3, \ldots$, and let $\bar{M}$ be the closure of M in the $L[0,1]$ norm. Since $1 \in \bar{M}$, the space $\bar{M}$ coincides with the closure of the space generated by $\{1, x^{n_p - n_1}\}$. From the fact that $\sum_{p=2}^{\infty} 1/(n_p - n_1) = \infty$ and the Müntz-Szász theorem we can conclude that $\bar{M} = L[0,1]$. From the Hahn-Banach theorem it follows that $\varphi_F \equiv 0$ in $L[0,1]$. Thus we have

$$\varphi_F(1) = \int_0^1 dF(x) = F(1) = \int_0^1 t^{n_1} f(t) dt = 0 = v,$$

which is a contradiction.$\|$

Corollary 6.2. Let $\{p_n(x)\}, n = 1, 2, ...$, be a sequence of polynomials (the degree of $p_n(x)$ is n) that are complete in $L[0, 1]$. If there exists a sequence of positive integers $\{n_j\}$ such that

$$\sum_{j=1}^{\infty} \frac{1}{n_j} = \infty \quad \text{and} \quad x^{n_j} = a_1(j)p_1(x) + \sum_{i=2}^{n_j} a_i(j)p_i(x),$$

where $a_1(j)$ is an integer for $j = 1, 2, ...$, then the subsequence $p_n(x), n = 2, 3, ...$, is also complete in $L[0, 1]$.

Proof. Let f be a function in $L[0, 1]$ satisfying

$$\int_0^1 f(x)p_m(x)dx = 0, \quad m = 2, 3, ... \ .$$

We then have

$$\int_0^1 x^{n_j} f(x)dx = a_1(j) \int_0^1 f(x)p_1(x)dx = a_1(j)v.$$

If $v \neq 0$ then, because of the restriction on $f(x)$, the set $\{a_1(j)\}, j = 1, 2, ...$, must be bounded. The assertion now follows from Theorem 6.3.‖

Theorem 6.4. Let ξ and η be random variables with distribution functions $F(x)$ and $G(x)$ respectively. If $E \mid \xi \mid < \infty$ and $E \mid \eta \mid < \infty$, then $F(x) = G(x)$ if and only if for every $n = 2, 3, ...$, there exists an integer $k_n : 1 \leq k_n \leq n$ such that

$$E\xi_{k_n,n} = E\eta_{k_n,n}.$$

Proof. It is easy to convince ourselves that the space generated by $\{x^n\}, n = 0, 1, ...$, coincides with the space generated by $\{p_{n-1}(x) = x^{k_n-1}(1-x)^{n-k_n}\}, n = 1, 2, ...$, so that the sequence $\{p_{n-1}(x)\}, n = 1, 2, ...$, is complete in $L[0, 1]$. It is sufficient to show that the subsequence $\{p_{n-1}(x)\}, n = 2, 3, ...$, is also complete in $L[0, 1]$. We shall prove by induction that all of the coefficients $a_0(n)$ are integers.

61

Let us assume that for all $1 \leq k < n$ we have

$$x^k = a_0(k)p_0(x) + \sum_{i=1}^{k} a_i(k)p_i(x),$$

where $p_0(x) = 1$, $a_0(k)$ is an integer. Then from the expansion

$$p_n(x) = x^{k_{n+1}-1} - C'_{n+1-k_{n+1}} x^{k_{n+1}} + \cdots + \delta x^n, \quad |\delta| = 1,$$

we get

$$x^n = a_0(n)p_0(x) + \sum_{i=1}^{n} C_i(n)p_i(x),$$

where $a_0(n) = -\delta \sum_{j=k_{n+1}-1}^{n-1} a_0(j), \ldots, a_n(n) = \delta$. So $a_0(n)$ is also an integer. Since $\sum_{n=1}^{\infty} \frac{1}{n} = \infty$ and $a_0(n)$ is an integer for $n = 1, 2, \ldots$, it follows from Corollary 6.2 that $\{p_{n-1}(x)\}, n = 2, 3, \ldots$, is also complete in $L[0, 1]$.

Since

$$\int_0^1 (F^{-1}(t) - G^{-1}(t))t^{k_n-1}(1-t)^{n-k_n} dt = \int_0^1 (F^{-1}(t) - G^{-1}(t))p_{n-1}(t)dt = 0$$

for $n = 2, 3, \ldots$, it follows from the completeness of $p_{n-1}(t)$ that $F(x) = G(x)$. $\|$

Theorem 6.5. Let ξ and η be random variables with distribution functions $F(x)$ and $G(x)$ respectively and assume that $E \mid \xi \mid < \infty$ and $E \mid \eta \mid < \infty$. In order that $F(x) = G(x)$, it is necessary and sufficient that there exist two sequences $\{m_j\}$ and $\{n_j\}$ of positive integers which satisfy:

$$E\xi_{k,n_j} = E\eta_{k,n_j}, \quad k = m_j, m_j + 1, \ldots, n_j,$$

and

$$\sum_{j=1}^{\infty} \sum_{k=m_j}^{n_j} \frac{1}{k} = \infty. \tag{6.4}$$

Proof. Let us prove that the family of polynomials

$$\{x^{k-1}(1-x)^{n_j-k}\}, \quad k = m_j, m_j + 1, \ldots, n_j \tag{6.5}$$

is complete in $L[0,1]$. Taking into account (3.10) it can be shown that the polynomials (6.5) are defined by the following family of polynomials:

$$\{x^k\}, \quad k = m_j, m + j + 1, \ldots, n_j. \tag{6.6}$$

It now follows from (6.4), (6.6) and the Müntz-Szász theorem that (6.5) is complete.‖

Corollary 6.3. Under the assumptions of Theorem 6.5, $F(x) = G(x)$ if and only if

$$E\xi_{k,n_j} = E\eta_{k,n_j}, \quad k = m, m+1, \ldots, n_j,$$

for some fixed m and the sequence $n_j \longrightarrow \infty$ as $j \longrightarrow \infty$.

Corollary 6.4. Under the assumptions of Theorem (6.5), $F(x) = G(x)$ if and only if there exists a sequence of integers $n_j, j = 1, 2, \ldots$, such that

$$E\xi_{n_j,n_j} = E\eta_{n_j,n_j}, \; j = 1, 2, \ldots, \quad \text{and} \quad \sum_{j=1}^{\infty} \frac{1}{n_j} = \infty.$$

With some additional restrictions on the sequence $\{k_n\}$ and on the distributions functions $F(x)$ and $G(x)$, S.T. Mkrtcjan [17] obtained a result which can be considered as a bound for the stability in Theorem 6.4.

Let us now look at a sequence $\{k_\ell\}$ that satisfies the condition

$$k_m \leq k_\ell \leq k_m + \ell - m$$

for all $\ell \geq m$. We introduce a class $\mathcal{F}$ of distribution functions satisfying the conditions:

63

(1) for any distribution $F(x) \in \mathcal{F}$, any $\epsilon > 0$, and any $t \in (0,1)$

$$\sum_{|x| \le \epsilon} | F^{-1}(x+t) - F^{-1}(t) | \le \rho_1(t, \epsilon),$$

where $\rho_1(t, \epsilon) \ge 0$ is some function defined for the whole class $\mathcal{F}$;

(2) for any $F(x) \in \mathcal{F}$

$$\gamma(m, F) = E | \xi_{k_m, m} | < \infty;$$

(3) for any $F(x) \in \mathcal{F}$ for all $t \in (0,1)$

$$| F^{-1}(t) | \le L(t),$$

where $L(t)$ is given and fixed for the whole class $\mathcal{F}$.

We set

$$\rho(t; \epsilon, m) = \rho_1(t; \epsilon) + (m-1)L(t).$$

Theorem 6.6. Let ξ and η be random variables with distribution functions $F(x)$ and $G(x)$ respectively. If $F(x) \in \mathcal{F}$ and $G(x) \in \mathcal{F}$ and

$$E\xi_{k_n, n} = E\eta_{k_n, n}$$

for $n = m, m+1, \ldots, m+s, (s \ge 1)$, then for any $\delta \in (0,1)$ and any $y \in (0,1)$,

$$y^{k_m - 1}(1-y)^{m-k_m} | F^{-1}(y) - G^{-1}(y) |$$
$$\le C \left[\frac{(k_m - 1)!(m - k_m)!}{m!} (\gamma(m; F) + \gamma(m; G)) \frac{1}{\delta^3} \frac{1}{\sqrt{s}} + \rho(y; \delta; m) \right],$$

where C is an absolute constant.

Let us study further the characterization of the exponential distribution using moments of higher order.

64

Theorem 6.7. The relation $\mu_{i+1,n}^{(2)} - \mu_{i,n}^{(2)} = 2(n-i)\lambda\mu_{i+1,n}^{(1)}$ for $i = 0, 1, ...; n = i+1, i+2, ...,$ holds if and only if $\bar{F}(x) = e^{-\lambda x}$.

Proof. Integrating (3.12) by parts we get

$$\mu_{i+1,n}^{(2)} = \mu_{i,n}^{(2)} + 2\frac{n!}{i!(n-i)!}\int_0^\infty xF^i(x)[1 - F(x)]^{n-i}dx, \tag{6.7}$$

that is, the characterization property is equivalent to

$$\int_0^\infty xF^i(x)[1 - F(x)]^{n-i}dx = \lambda \int_0^\infty xF^i(x)[1 - F(x)]^{n-i-1}dF(x)$$

or

$$\int_0^1 u^i(1-u)^{n-i-1}F^{-1}(u)\{(1-u)(F^{-1}(u))' - \lambda\}du = 0$$

for $i = 0, 1, \ldots, ; n = i+1, i+2, \ldots.$

A continuous function, orthogonal to $u^i(1-u)^{n-1-i}$ for $n = i+1, i+2, ...,$ is zero. Therefore $(1-u)(F^{-1}(u))' = \lambda$ for almost all $u(0 < u < 1)$, which implies $\bar{F}(x) = e^{-\lambda x}$. The converse assertion is trivial.‖

Theorem 6.8. The relation $n(\mu_{i,n}^{(2)} - \mu_{i-1,n-1}^{(1)}) = 2\lambda\mu_{i,n}^{(1)}$ for $i = 1, 2, \ldots, ;$ $n = i, i+1, ...,$ holds if and only if $\bar{F}(x) = e^{-\lambda x}$.

Proof. The case $i = 1$. Integrating (3.12) by parts we get

$$\mu_{1,n}^{(2)} = 2\int_0^\infty x[1 - F(x)]^n dx.$$

Therefore the characterization property is equivalent to

$$\int_0^\infty [1 - F(x)]^n dx = \lambda \int_0^\infty x[1 - F(x)]^{n-1}dF(x)$$

or

$$\int_0^1 F^{-1}(u)(1-u)^{n-1}\{(1-u)(F^{-1}(u))' - \lambda\}du = 0 \quad n = 1, 2, \ldots.$$

65

Therefore $(F^{-1}(u))' = \lambda(1-u)^{-1}$ for almost all $u(0 < u < 1)$, hence $\bar{F}(x) = e^{-\lambda x}$.

The case $i > 1$. Using the recurrence relation (3.17) we get

$$n(\mu_{i+1,n}^{(2)} - \mu_{i,n-1}^{(2)}) = (n-i)(\mu_{i+1,n}^{(2)} - \mu_{i,n}^{(2)}),$$

and the proof follows from Theorem 6.6.$\|$

Theorem 6.9. The relation

$$\sigma_{i+1,i+1,n} + \sigma_{i,i,n} = (\mu_{i+1,n}^{(1)} - \mu_{i,n}^{(1)})^2 \text{ for } i = 0, 1, \ldots, ; \ n = i+1, i+2, \ldots,$$

holds if and only if $\bar{F}(x) = e^{-\lambda x}$.

Proof. The condition of the theorem can be rewritten as follows

$$(\mu_{i+1,n}^{(2)} - \mu_{i,n}^{(2)})/(\mu_{i+1,n}^{(1)} - \mu_{i,n}^{(1)}) = 2\mu_{i+1,n}^{(1)}. \tag{6.8}$$

Integrating by parts we get

$$\mu_{i+1,n}^{(1)} - \mu_{i,n}^{(1)} = \frac{n!}{i!(n-i-1)!} \int_0^\infty F^i(x)[1 - F(x)]^{n-i} dx. \tag{6.9}$$

Substituting (6.7) and (6.9) in (6.8) leads to

$$\frac{\int_0^\infty x F^i(x)[1 - F(x)]^{n-i} dx}{\int_0^\infty F^i(x)[1 - F(x)]^{n-i} dx} = \frac{\int_0^\infty x F^i(x)[1 - F(x)]^{n-i-1} dF(x)}{\int_0^\infty F^i(x)[1 - F(x)]^{n-i-1} dF(x)}.$$

So, we have

$$\int_0^\infty x F^i(x)[1 - F(x)]^{n-i} dx = \lambda \int_0^\infty x F^i(x)[1 - F(x)]^{n-i-1} dF(x)$$

and

$$\int_0^\infty F^i(x)[1 - F(x)]^{n-i} dx = \int_0^\infty F^i(x)[1 - F(x)]^{n-i-1} dF(x)$$

for some $\lambda > 0$, or

$$\int_0^1 u^i (1-u)^{n-i-1} F^{-1}(u)[(F^{-1}(u))'(1-u) - \lambda] du = 0.$$

Consequently, $(F^{-1}(u))' = \lambda(1-u)^{-1}$ for almost all $0 < u < 1$, and $\bar{F}(x) = \exp\{-\lambda x\}$. $\|$

Chapter 7
CHARACTERIZATION PROBLEMS OF THE GEOMETRIC DISTRIBUTION

We know that the geometric distribution is the discrete analogue of the exponential distribution. It is natural to expect that characterization properties of the geometric distribution will be discrete analogues of the characterization properties of the exponential distribution as well. We illustrate this fact by some examples.

We shall say that an integer-valued random variable ξ has a geometric distribution if

$$P\{\xi = k\} = p(1 - p)^{k-1}, k = 1, 2, \cdots, \quad 0 < p < 1.$$

Let us look at an example in which the geometric distribution arises. Consider an experiment where the probability is p that the event A occurs during the course of the experiment. If independent experiments are carried out successively, then the probability that event A occurs for the first time in the k-th experiment is equal to $p(1 - p)^{k-1}$, that is, the distribution will be geometric.

It is not difficult to show that the geometric distribution also has the lack of memory property. Let the random variable ξ be the waiting time of the event A with ξ taking values $1, 2, \ldots$ with probabilities $p_1, p_2, \ldots$.

Theorem 7.1. If the conditional probability that the waiting time is terminated on the k-th trial given that it did not terminate earlier is equal to p_1, that is,

$$P\{\xi = k \mid \xi > k - 1\} = P\{\xi = 1\}, \tag{7.1}$$

then $p_k = (1 - p_1)^{k-1} p_1$ so that ξ has a geometric distribution.

Proof. Let

$$q_k = p_{k+1} + p_{k+2} + \cdots = P\{\xi > k\}.$$

Then (7.1) can be written in the form

$$\frac{p_k}{q_{k-1}} = p_1,$$

and since $p_k = q_{k-1} - q_k$,

$$\frac{q_k}{q_{k-1}} = 1 - p_1.$$

Because $q_1 = p_2 + p_3 + \cdots = 1 - p_1$, we have $q_k = (1 - p_1)^k$ and hence $p_k = q_{k-1} - q_k = (1 - p_1)^{k-1}p_1$, which was to be proved.||

Theorem 7.2. Let $\xi_1, \xi_2, \ldots$ be independent, identically distributed, positive integer-valued random variables. The conditional distribution of the random variable $R = \xi_{2,2} - \xi_{1,2}$, under the condition that $R > 0$, coincides with the distribution of ξ_1 if and only if $P\{\xi_1 = k\} = p(1 - p)^{k-1}, k = 1, 2, \ldots$, for some $0 < p < 1$.

Proof. Since $p_k = P\{\xi_1 = k\}$ we have

$$P\{R = r \mid R > 0\} = c \sum_{k=1}^{\infty} p_k p_{r+k}, \quad r = 1, 2, \ldots,$$

where $c = 2/P\{R > 0\}$.

If the distribution is geometric, the assertion of the theorem is obtained by computation. If $P\{R = r \mid R > 0\} = P\{\xi_1 = r\}$, then

$$p_r = c \sum_{k=1}^{\infty} p_k p_{r+k}, \quad r = 1, 2, \ldots.$$

Hence, we immediately obtain the inequalities

$$p_r \geq c p_1 p_{r+1}, \quad r = 1, 2, \ldots; \tag{7.2}$$

69

$$p_{r+1} = c \sum_{k=1}^{\infty} p_k p_{r+1+k} \geq c^2 p_1 \sum_{k=1}^{\infty} p_{k+1} p_{k+1+r} = c^2 p_1 [c^{-1} p_r - p_1 p_{r+1}].$$

Thus,

$$p_{r+1} \geq [c p_1 / (1 + c^2 p_1^2)] p_r, \quad r = 1, 2, \ldots . \tag{7.3}$$

Now define

$$\beta = \sup\{b; b > 0 : p_r \geq b p_{r+1} \text{ for all } r = 1, 2, \ldots\} \tag{7.4}$$

$$\gamma = \sup\{d; 0 < d < 1 : p_{r+1} \geq p_r d \text{ for all } r = 1, 2, \ldots\}. \tag{7.5}$$

The existence of such β and γ is guaranteed by formulas (7.2) and (7.3). In addition,

$$p_r \geq \beta p_{r+1} \geq \beta \gamma p_r, \text{ that is } \beta \gamma \leq 1.$$

If we can show that $\beta \gamma \geq 1$, then it will follow that $\beta \gamma = 1$ and consequently, $p_{r+1} = \gamma p_r$ for some $0 < \gamma < 1$ and for all $r = 1, 2, \ldots$. The theorem holds for $p_1 = p$ and $\gamma = 1 - p$.

In order to prove that $\beta \gamma \geq 1$, note that it follows from (7.4) that

$$p_{r+1} = c \sum_{k=1}^{\infty} p_k p_{r+1+k} \geq c \beta \sum_{k=1}^{\infty} p_{k+1} p_{k+1+r} = c \beta [c^{-1} p_r - p_1 p_{r+1}], \quad r = 1, 2, \ldots .$$

Hence, taking into account (7.5) we have

$$\gamma \geq \frac{\beta}{1 + c \beta p_1}, \text{ or } \beta - \gamma \leq c p_1 \beta \gamma. \tag{7.6}$$

Furthermore,

$$p_r \geq c p_1 p_{r+1} + c \gamma \sum_{k=2}^{\infty} p_{k-1} p_{k+r} = c p_1 p_{r+1} + c \gamma \sum_{k=1}^{\infty} p_k p_{r+1+k}$$

$$= (c p_1 + \gamma) p_{r+1}, \quad r = 1, 2, \ldots .$$

Using (7.4) we get

$$\beta \geq c p_1 + \gamma, \text{ or } \beta - \gamma \geq c p_1. \tag{7.7}$$

70

From (7.6) and (7.7) it follows that $\beta\gamma \geq 1$.

Theorem 7.3. Let ξ be a non-degenerate discrete random variable for which the set of possible values can be represented as a strictly increasing sequence of real numbers $\{a_i, \ i \in I\}$.

Let k be an arbitrary fixed positive integer, $(2 \leq k \leq n)$. Then $\xi_{1,n}$ is independent of the event $\{\xi_{k,n} = \xi_{1,n}\}$ if and only if

$$P\{\xi \geq a_i\} = q^{i-1}, \ \ i = 1, 2, \dots, \ \text{ where } \ 0 < q < 1.$$

Proof. If $P\{\xi \geq a_i\} = q^{i-1}$, then the independence of $\xi_{1,n}$ and $\{\xi_{k,n} = \xi_{1,n}\}$ is obvious. Let us prove the converse. Let $\bar{F}(i) = P\{\xi \geq a_i\}$. Then, by assumption we have

$$p\{\xi_{k,n} = \xi_{1,n}, \ \xi_{1,n} = a_i\} = P\{\xi_{k,n} = \xi_{1,n}\}P\{\xi_{1,n} = a_i\}.$$

Rewriting, we get

$$P\{\xi_{1,n} = \xi_{k,n} = a_i\} = \sum_{j=k}^{n} \binom{n}{j} [\bar{F}(i) - \bar{F}(i+1)]^j \bar{F}^{n-j}(i+1),$$

and setting $j' = n - j$, we have

$$\sum_{j'=0}^{n-k} \binom{n}{j'} \bar{F}^{j'}(i+1)[\bar{F}(i) - \bar{F}(i+1)]^{n-j'} = P\{\xi_{1,n} = \xi_{k,n}\}[\bar{F}^n(i) - \bar{F}^n(i+1)] \ \text{ for all } i \in I. \ \ (7.8)$$

Furthermore, either $I = \{i \in N : i \leq r\}$ for some $r \in N$ or $I = N$. In the case $I = \{i \in N : i \leq r\}$ for some $r \in N$, let us set $i = r$ in (7.8); then

$$\bar{F}^n(r) = P\{\xi_{1,n} = \xi_{k,n}\}\bar{F}^n(r), \ \text{ where } \ \bar{F}(r) > 0.$$

Therefore, we must have $P\{\xi_{1,n} = \xi_{k,n}\} = 1$ which is possible only if ξ is degenerate.

Now assume that $I = N$. Dividing both sides of (7.8) by $\bar{F}^n(i)$ and setting $q(i) =$

71

$\bar{F}(i+1)/\bar{F}(i)$ we get

$$\sum_{j=0}^{n-k} \binom{n}{j} q^j(i)[1-q(i)]^{n-j}(1-q^n(i))^{-1} = P\{\xi_{1,n} = \xi_{k,n}\} \tag{7.9}$$

for $i = 1, 2, \dots$.

Note that $0 < q(i) < 1$. Let η_i be a r.v. having a binomial distribution with parameters $(n, q(i))$, $i = 1, 2, \dots$. Then the numerator of the left-hand side of (7.9) is $P\{\eta_i \leq n - k\}$. Since

$$P\{\eta_i \leq n-k\} = 1 - P\{\eta_i \geq n-k+1\} = k\binom{n}{k}\int_{q(i)}^{1} u^{n-k}(1-u)^{k-1}du,$$

the left-hand side of (7.9) can be written as

$$\{k\binom{n}{k}\int_{0}^{1-q(i)} t^{k-1}(1-t)^{n-k}dt\}/(1-q^n(i)).$$

And since the right-hand side of (7.9) does not depend on i, the left- hand side is constant for $i = 1, 2, 3, \dots$. Set

$$f(x) = \{k\binom{n}{k}\int_{0}^{1-x} t^{k-1}(1-t)^{n-k}dt\}/(1-x^n), \quad 0 < x < 1.$$

Differentiating with respect to x we get

$$f'(x) = \{k\binom{n}{k}x^{n-k}[nx^{k-1}\int_{0}^{1-x} t^{k-1}(1-t)^{n-k}dt - (1-x)^{k-1}(1-x^n)]\}/(1-x^n)^2.$$

To prove that $f'(x) < 0, 0 < x < 1$, note that

$$nx^{k-1}\int_{0}^{1-x} t^{k-1}(1-t)^{n-k}dt - (1-x)^{k-1}(1-x^n)$$

$$\leq (1-x)^{k-1}[nx^{k-1}\int_{0}^{1-x}(1-t)^{n-k}dt - (1-x^n)]$$

$$= \frac{(1-x)^{k-1}}{n-k+1}[nx^{k-1} - (n-k+1) - (k-1)x^n].$$

Also, set $g(x) = nx^{k-1} - (n - k + 1) - (k - 1)x^n$. Since $g(0) < 0, g(1) = 0$ and $g'(x) = n(k - 1)x^{k-2}(1 - x^{n-k+1}) > 0$ for $0 < x < 1$, it follows that $g(x) < 0$ for $0 < x < 1$.

Therefore, $f'(x) < 0, 0 < x < 1$, which implies that $f(x)$ is strictly decreasing. Taking into account (7.9) we find that $q(i)$ is constant for $i = 1, 2, \ldots$. Set $q(i) = q, 0 < q < 1$. Then it follows that $\bar{F}(i) = q^{i-1}$, and the theorem is proved.$\|$

Corollary 7.1. Assume that ξ satisfies the conditions of Theorem 7.3. The random variable $\xi_{1,n}$ is independent of the event $\{\xi_{k,n} > \xi_{1,n}\}$ if and only if

$$P\{\xi \geq a_i\} = q^{i-1}, \quad i = 1, 2, \ldots, \quad 0 < q < 1.$$

The proof follows quickly from the fact that the event $\{\xi_{k,n} > \xi_{1,n}\}$ is the complement of the event $\{\xi_{k,n} = \xi_{1,n}\}$.

We now proceed to some theorems on the stability of the geometric distribution. Define a failure rate function

$$r(n) = \frac{p_n}{\sum_{i=n}^{\infty} p_i} \tag{7.10}$$

for a discrete distribution and classes of discrete IFR-and DFR-distributions depending on whether $r(n)$ increases or decreases with increasing n. Obviously $r(n)$ will be defined only for those n for which $\sum_{i=n}^{\infty} p_i \neq 0$. The geometric distribution will be characterized by $r(n)$ constant. If a discrete distribution has a monotone failure rate then, just as for the continuous case, we shall say that it belongs to class C.

Theorem 7.4. If $\{p_n\}_{n=1}^{\infty} \in C$ then we have the bound

$$\sup_{n \geq 1} |\sum_{k=n}^{\infty} p_k - q^{n-1}| \leq |1 - \mu_1 r(1)|,$$

where $\mu_i = \sum_{k=1}^{\infty} k^i p_k$, $q = (\mu_1 - 1)/\mu_1$.

Theorem 7.5. If $\{p_n\}_{n=1}^{\infty} \in C$ then we have the bound

$$\sum_{n \geq 1} | \sum_{k=n}^{\infty} p_k - q^{n-1} | \leq \sqrt{(1 - \mu_1 r(1))(1 - \frac{\sigma^2 + \mu_1}{\mu_1^2})},$$

where $\sigma^2 = \mu_2 - \mu_1^2$ and μ_i and q are defined as in Theorem 7.4.

Corollary 7.2. If $\{p_n\}_{n=1}^{\infty} \in C$ then $\{p_n\}$ is a geometric distribution if and only if either $\mu_1 r(1) = 1$ or $\mu_2 = 2\mu_1^2 - \mu_1$.

Proofs of Theorems 7.4 and 7.5. We consider only the case of a non-decreasing $r(n)$.

Let η be a discrete random variable with distribution $\{\tilde{p}_n\}_{n=1}^{\infty}, \tilde{p}_n = q_n/\mu_1$, where $q_n = p_n + p_{n+1} + \dots$, and $n \geq 0$. The failure rate function for the distribution $\{\tilde{p}_n\}$ is $\tilde{r}(n) = q_n / \sum_{k=n}^{\infty} q_k$ which also does not decrease with n. Besides, $\tilde{r}(n) \geq r(n)$ which gives $p_1 \mu_1 \leq 1$ for $n = 1$.

Consider the function

$$\varphi(n) = -\frac{1}{\mu_1} \sum_{k=n}^{\infty} q_k + q_n. \tag{7.11}$$

Since $\tilde{r}(1) = 1/\mu_1$, $\varphi(n)$ is non-negative:

$$\varphi(n) = \sum_{k=n}^{\infty} q_k(\tilde{r}(n) - \frac{1}{\mu_1}) \geq \sum_{k=n}^{\infty} q_k(\tilde{r}(1) - \frac{1}{\mu_1}) = 0.$$

Set $f(n) = \sum_{k=r}^{\infty} q_k$ and use the notation $\Delta f(n) = f(n+1) - f(n) = -q_n$. Then (7.11) can be written as

$$\varphi(n) = -\Delta f(n) - \frac{1}{\mu_1} f(n). \tag{7.12}$$

If $f(n)$ is considered as an unknown function, then (7.12) is a linear difference equation. Its general solution is

$$f(n) = (\frac{\mu_1 - 1}{\mu_1})^{n-1} [-\sum_{k=1}^{n-1} \varphi(k)(\frac{\mu_1}{\mu_1 - 1})^k + f(1)].$$

74

Since $f(1) = \mu_1$, the solution has the form

$$f(n) = - \sum_{k=1}^{n-1} \varphi(k) q^{n-1-k} + q^{n-1} \mu_1, \qquad (7.13)$$

where $q = (\mu_1 - 1)/\mu_1$. Rewrite (7.11) using (7.13):

$$q_n - q^{n-1} = \varphi(n) - \frac{1}{\mu_1} \sum_{k=1}^{n-1} \varphi(k) q^{n-1-k}. \qquad (7.14)$$

Since $\varphi(k) \geq 0$, we have from (7.14) that

$$q_n - q^{n-1} \leq \varphi(n). \qquad (7.15)$$

Consider

$$\Delta_n \equiv \varphi(n+1) - \varphi(n) = q_{n+1} - q_n + \frac{q_n}{\mu_1} = q_n \left(\frac{1}{\mu_1} - r(n) \right). \qquad (7.16)$$

Since $r(n)$ does not decrease and $r(1) = p_1 \leq \tilde{r}(1) = 1/\mu_1$, it follows from (7.16) that Δ_n is positive at first and negative after than. Let m be a unique number for which $r(m) \leq 1/\mu_1$ and $r(m+1) > 1/\mu_1$. Then $\max_n \varphi(r) = \varphi(m)$ and, using (7.15), we get

$$q_n - q^{n-1} \leq \varphi(m) \text{ for all } n = 1, 2, \dots . \qquad (7.17)$$

On the other hand it follows from (7.14) that

$$q_n - q^{n-1} \geq -\frac{1}{\mu_1} \sum_{k=1}^{n-1} \varphi(k) q^{n-1-k} \geq -\varphi(m) \frac{1}{\mu_1(1-q)} = -\varphi(m). \qquad (7.18)$$

Inequalities (7.17) and (7.18) yield

$$\mid q_n - q^{n-1} \mid \leq \varphi(m). \qquad (7.19)$$

Obviously

$$\varphi(m) = \sum_{k=1}^{m-1} \Delta_k \leq \sum_{k=1}^{m-1} q_k \left(\frac{1}{\mu_1} - r(1) \right) \leq \sum_{k=1}^{\infty} q_k \frac{1}{\mu_1} (1 - \mu_1 r(1)) = 1 - \mu_1 r(1).$$

The bound, together with (7.19), prove Theorem 7.4.

Furthermore,

$$\varphi(m) \leq \sum_{k=1}^{m-1} q_k(\frac{1}{\mu_1} - r(1)) \leq \frac{m}{\mu_1}(1 - \mu_1 r(1)),$$

and since $\{\varphi(n)\}$ for $1 \leq n \leq m$ is convex upwards,

$$\tilde{\varphi}(\infty) = \sum_{k=1}^{\infty} \varphi(k) \geq \sum_{k=1}^{m} \varphi(k) \geq \frac{m}{2}\varphi(m) \geq \frac{\mu_1}{2}\varphi^2(m)/(1 - r(1)\mu_1),$$

so that

$$\varphi(m) \leq \sqrt{\frac{2\tilde{\varphi}(\infty)}{\mu_1}}(1 - \mu_1 r(1)).$$

Taking into account that $\tilde{\varphi}(\infty) = \mu_1(1 - \frac{\mu_1 - \mu_1}{2\mu_1^2})$ and using (7.19) we obtain the assertion of Theorem 7.5.$\|$

The proof of Corollary 7.2 follows immediately from the bounds in Theorem 7.5.

Theorem 7.6. Let ξ be a non-degenerate random variable with $P\{\xi = k\} = p_k$, $k = 1, 2, \dots$. The relation

$$D\{\xi \mid \xi \geq k\} = d = \text{ constant}, \quad k = 1, 2, \dots, \tag{7.20}$$

holds if and only if ξ has a geometric distribution.

Proof. If ξ has a geometric distribution, then the conditional distribution of ξ under the condition that $\xi \geq k$ coincides with the distribution of the random variable $\xi + k$. Therefore $D\{\xi \mid \xi \geq k\} = D\xi$ for all k.

To prove the reverse assume that (7.20) is satisfied. Then

$$D\{\xi \mid \xi \geq k\} = E\{(\xi - k)^2 \mid \xi \geq k\} - [E(\xi - k) \mid \xi \geq k)]^2. \tag{7.21}$$

Let $F_k(x) = P\{\xi < x \mid \xi \geq k\}$, $\bar{F}_k(x) = 1 - F_k(x)$. By definition

$$E\{(\xi - k)^2 \mid \xi \geq k\} = \int_k^{\infty} (x - k)^2 dF_k(x).$$

76

Integration by parts yields

$$E\{(\xi - k)^2 \mid \xi \geq k\} = \frac{\sum_{i=k}^{\infty} p_i(i - k)^2}{\sum_{i=k}^{\infty} p_i} = \sum_{i=k+1}^{\infty} \frac{q_i}{q_k}[2(i - k) - 1]. \tag{7.22}$$

Thus we get

$$E\{\xi - k \mid \xi \geq k\} = \sum_{i=k+1}^{\infty} \frac{q_i}{q_k}. \tag{7.23}$$

Substituting (7.21)-(7.23) in (7.20) we obtain

$$q_k \sum_{i=k+1}^{\infty} q_i[2(i - k) - 1] - [\sum_{i=k+1}^{\infty} q_i]^2 = dq_k^2. \tag{7.24}$$

This equation still holds if k is replaced by $k + 1$. Subtracting the equation obtained in this way from (7.24) and simplifying, we have

$$\sum_{i=k+1}^{\infty} q_i[2(i - k) - 1] = d(q_k + q_{k+1}). \tag{7.25}$$

Substituting $k + 1$ for k in (7.25) and subtracting,

$$q_{k+1} + 2 \sum_{i=k+2}^{\infty} q_i = d(q_k - q_{k+2}).$$

Repeating this procedure once again, we obtain the linear difference equation

$$dq_{k+3} - (1 + d)q_{k+2} - (1 + d)q_{k+1} + dq_k = 0. \tag{7.26}$$

Its characteristic equation becomes

$$dx^3 - (d + 1)x^2 - (d + 1)x + d = 0$$

whose roots are: $x_1 = -1$, $x_2 = [2d + 1 - \sqrt{4d + 1}]/2d$ and $x_3 = [2d + 1 + \sqrt{4d + 1}]/2d$. The roots are all real for $d > 0$. It can be shown that $0 < x_2 < 1 < x_3$. The general solution of (7.26) has the form

$$q_k = C_0(-1)^k + C_1 x_2^k + C_2 x_3^k.$$

Since $\lim_{k \to \infty} q + k = 0$, we must have $C_0 = C_2 = 0$. Thus, $q_k = C_1 x_2^k$. From the initial condition $q_1 = 1$ we have $q_k = x_2^{k-1}$ with $0 < x_2 < 1$. Hence it follows that the distribution is geometric.$\|$

Chapter 8
CHARACTERIZATION OF THE EXPONENTIAL DISTRIBUTION USING THE GEOMETRIC DISTRIBUTION

Let $\{\xi_i\} \in P$ be a sequence of independent identically distributed random variables and let ν be a random variable having the geometric distribution

$$P\{\nu = k\} = \epsilon(1 - \epsilon)^{k-1}, \quad k = 1, 2, \ldots, \ 0 < \epsilon < 1,$$

and not depending on the sequence of random variables $\{\xi_i\}$.

Theorem 8.1. The relation

$$P\{\epsilon(\xi_1 + \cdots + \xi_\nu) \geq x\} = P\{\xi_1 \geq x\}, \tag{8.1}$$

is satisfied for all $x \geq 0$ if and only if $\xi_1 \in \mathcal{E}$.

Proof. Let $f(s) = Ee^{-s\xi_1}, \varphi(s) = Ee^{s\epsilon(\xi_1 + \cdots + \xi_\nu)}$, Re $s \geq 0$. Then it follows from (8.1) that

$$\varphi(s) \equiv f(s).$$

But, on the other hand

$$\varphi(s) = \sum_{k=1}^{\infty} P\{\nu = k\} E \exp -s\epsilon(\xi_1 + \cdots + \xi_k)$$

$$= \sum_{k=1}^{\infty} \epsilon(1 - \epsilon)^{k-1} f^k(s\epsilon) = \frac{\epsilon f(s\epsilon)}{1 - (1 - \epsilon)f(s\epsilon)}.$$

Thus, if (8.1) is satisfied, then for all Re $s \geq 0$ the identity

$$f(s) = \frac{\epsilon f(\epsilon s)}{1 - (1 - \epsilon)f(\epsilon s)}$$

holds. Iterating this equality we find that

$$f(s) = \frac{\epsilon^n f(\epsilon^n s)}{1 - (1 - \epsilon^n) f(\epsilon^n s)}$$

for all $n = 1, 2, \dots$. Hence

$$f(\epsilon^n s) = \frac{f(s)}{\epsilon^n + (1 - \epsilon^n) f(s)}. \tag{8.2}$$

Using this equality we show that for all Re $s \geq 0$

$$f(s) \neq 0, \quad f(s) \neq 1.$$

Indeed, let there exist some $s = s_0$ for which $f(s_0) = 0$. Then, from (8.2) we get $f(s_0 \epsilon^n) = 0$ for all $n \geq 1$. But, from the continuity of $f(s)$, $f(0) = 0$ which contradicts the property $f(0) = 1$ of the Laplace-Stieltjes transformation of a probability distribution. Likewise, if $f(s_0) = 1$ for some $s = s_0$, then it follows from (8.2) that

$$f(s_0 \epsilon^n) = 1, \quad n = 1, 2, \dots .$$

Hence, $f(s) \equiv 1$, that is, $P\{\xi_1 = 0\} = 1$ which contradicts the condition of the theorem.

Let us rewrite (8.2) in the following form:

$$\frac{1 - f(s)}{s f(s)} = \frac{1}{f(s \epsilon^n)} \frac{1 - f(\epsilon^n s)}{s \epsilon^n}; \quad s \neq 0.$$

Therefore, as $n \to \infty$ the limit of the right-hand side exists and is equal to the mathematical expectation of the random variable ξ_1.

Let $E\xi_1 = 1/\lambda$; then, for all $s \neq 0$, we have

$$\frac{1 - f(s)}{s f(s)} = \frac{1}{\lambda},$$

or

$$f(s) = \frac{\lambda}{\lambda + s}.$$

For $s = 0$ we determine $f(s)$ by continuity. Since the Laplace-Stieltjes transformation uniquely defines the distribution function, we have $\xi_1 \in \mathcal{E}$. ||

Let $\xi \in \mathcal{P}$ be a random variable with distribution function $F(x)$ where $0 < F(x) < 1$ for $x > 0$. We specify some number $t > 0$ and look at the discrete random variable ξ_t which takes values $1, 2, \ldots, n, \ldots$ with probabilities

$$p_t(k) = P\{\xi_t = k\} = P\{\xi \in [(k-1)t, kt]\} = F(kt) - F((k-1)t). \tag{8.3}$$

Theorem 8.2. Let $\ell > 2$ be some natural number. Then $F(x)$ is an exponential distribution function if and only if for any $t > 0$ the probability $p_t(k)$ coincides with the corresponding geometric probability for $k = 1, 2, \ell$.

Corollary 8.1. The distribution function $F(x)$ is exponential with parameter $\lambda > 0$ if and only if $p_t(k)$ is a geometric distribution for any $t > 0$.

We shall use the following lemma to prove Theorem 8.2.

Lemma 8.1. Let $g : (0, \infty) \longrightarrow (0, 1)$ satisfy

$$g(2t) = g^2(t). \tag{8.4}$$

Then

$$g(t) = \exp\{-t2^{C \log_2 t}\} \tag{8.5}$$

where $C(\cdot)$ is a periodic function with period 1.

Proof. Let $A(t) = \log(-\log g(t))$. Then (8.4) can be rewritten as

$$A(2t) = A(t) + \log 2.$$

Using the notation $s = \log_2 t$, $B(s) = A(2^s)/\log 2$, we get

$$B(s+1) = B(s) + 1.$$

81

This is a difference equation with continuous argument and its general solution is

$$B(s) = s + C(s)$$

where $C(s)$ is an arbitrary periodic function with period 1. The expression (8.5) follows from this.‖

Proof of Theorem 8.2. We proceed immediately to the proof of sufficiency since the necessity is obvious. From the condition of the theorem it follows that

$$p_t(2) = p_t(1)(1 - p_t(1)) \tag{8.6}$$

and

$$p_t(\ell) = p_t(1)(1 - p_t(1))^{\ell-1}. \tag{8.7}$$

Using the definition of $p_t(k)$, we have from (8.6) and (8.7)

$$F(2t) - F(t) = F(t)(1 - F(t)) \tag{8.6'}$$

and

$$F(\ell t) - F((\ell - 1)t) = F(t)(1 - F(t))^{\ell-1}. \tag{8.7'}$$

From (8.6') it follows that

$$\bar{F}(2t) = \bar{F}^2(t).$$

Since $0 < \bar{F}(t) < 1$, applying Lemma 8.1

$$F(t) = 1 - \exp\{-t \cdot 2^{C(\log_2 t)}\}, \tag{8.8}$$

where $C(\cdot)$ is a periodic function with period 1. Using (8.8) let us rewrite (8.7') as

$$\exp\{-(\ell - 1)t \cdot 2^{C(\log_2 t + a)}\} - \exp\{-\ell t \cdot 2^{C(\log_2 t + b)}\}$$
$$= \exp\{-(\ell - 1)t \cdot 2^{C(\log_2 t)}\}\{1 - \exp\{-t \, 2^{C(\log_2 t)}\}\}, \tag{8.9}$$

where $a = \log_2(\ell - 1)$, $b = \log_2 \ell$.

For any natural n and real t_0 and d we have the identity

$$C(\log_2 2^{-n} t_0 + d) = C(\log_2 t_0 + d).$$

Setting $t = t_0 2^{-n}$, expanding (8.9) in a Taylor series to first order terms, multiplying by 2^n, and letting $n \to \infty$, we obtain

$$\ell \cdot 2^{C(\log_2 t_0 + b)} - (\ell - 1) 2^{C(\log_2 t_0 + a)} = 2^{C(\log_2 t_0)}. \tag{8.10}$$

Substituting $C(\log_2 t_0 + b)$ from (8.10) in (8.9), we find

$$C(\log_2 t_0 + a) = C(\log_2 t_0).$$

Therefore, $C(\cdot)$ has periods 1 and a. Let us look at the two cases:

(a) $a = \log_2(\ell - 1)$ is irrational. Then $C(\cdot)$ has periods of the form $m + na$ where m and n are any integers. And since the set $\{m + na\}$ is everywhere dense on $(-\infty, +\infty)$, $F(t) = 1 - e^{-\lambda t}$;

(b) a is rational, that is, $\ell - 1 = 2^r$ for some natural r. Then (8.9) is equivalent to

$$C(\log_2 t + b) = C(\log_2 t).$$

Then b will be irrational and $F(t) = 1 - e^{\lambda t}$. $\|$

The proof of Corollary 8.1 is obvious.

Let us now consider the stability of the characterization property formulated in Corollary 8.1.

Theorem 8.3. If $\lim\limits_{x \to +0} F(x)/x = \lambda < \infty$ and for some $0 < \epsilon < \lambda/2$ and all $t > 0$ there exists an $a = a(t)$ such that

$$\sup_{k \geq 1} \left| \frac{P\{\xi_t \geq k\} - a^{k-1}(t)}{k} \right| \leq \epsilon t, \tag{8.11}$$

then

$$\sup_{x \geq 0} \left| \frac{\bar{F}(x) - e^{-\lambda x}}{x} \right| < 4\epsilon.$$

Proof. From (8.3) and (8.11) we have

$$\sup_{k \geq 1} \left| \frac{\bar{F}((k-1)t) - a^{k-1}(t)}{k} \right| \leq \epsilon t.$$

Therefore the equality

$$\bar{F}((k-1)t) = a^{k-1}(t) + R_\epsilon(k;t) \tag{8.12}$$

holds, where

$$\mid R_\epsilon(k;t) \mid \leq \epsilon k t$$

for all $t > 0$ and $k = 1, 2, \dots$.

From (8.12) we obtain, for $k = 2$,

$$a(t) = \bar{F}(t) + R_\epsilon(2;t).$$

Rewrite (8.12) in the form:

$$\bar{F}(kt) = [\bar{F}(t) + R_\epsilon(2;t)]^k + R_\epsilon(k+1;t).$$

Setting $t = \tau/k$ we find

$$\bar{F}(\tau) = [\bar{F}(\tau/k)]^k \left[1 + \frac{R_\epsilon(2;\tau/k)}{\bar{F}(\tau/k)}\right]^k + R_\epsilon(k+1;\tau/k). \tag{8.13}$$

It is easy to see that

$$\limsup_{k \to \infty} \left[1 + \frac{R_\epsilon(2;\tau/k)}{\bar{F}(\tau/k)}\right]^k \leq \limsup_{k \to \infty} \left[1 + \frac{\mid R_\epsilon(2;\tau/k) \mid}{\bar{F}(\tau/k)}\right]^k$$

$$\leq \lim_{k \to \infty} [1 + \frac{2\epsilon\tau}{k}]^k = e^{2\epsilon\tau} \tag{8.14}$$

84

and

$$\liminf_{k \to \infty} \left[1 + \frac{R_\epsilon(2; \tau/k)}{\bar{F}(\tau/k)} \right]^k \geq \lim_{k \to \infty} [1 - \frac{2\epsilon\tau}{k}]^k = e^{-2\epsilon\tau}. \tag{8.15}$$

In addition

$$\mid R_\epsilon(k + 1; \frac{\tau}{k}) \mid \leq \frac{\epsilon(k+1)\tau}{k} \leq 2\epsilon\tau \tag{8.16}$$

and

$$\lim_{k \to \infty} [\bar{F}(\frac{\tau}{k})]^k = \exp\{-\lambda\tau\}. \tag{8.17}$$

Thus, from relations (8.13)–(8.17) it follows that

$$e^{-\lambda\tau} e^{-2\epsilon\tau} - 2\epsilon\tau \leq \bar{F}(\tau) \leq e^{-\lambda\tau} e^{2\epsilon\tau} + 2\epsilon\tau.$$

From this we immediately get the required inequality

$$\left| \frac{\bar{F}(\tau) - e^{-\lambda\tau}}{\tau} \right| \leq 2\epsilon + \frac{e^{2\epsilon\tau} - 1}{\tau} e^{-\lambda\tau} \leq 4\epsilon. \parallel$$

Chapter 9

MULTIVARIATE EXPONENTIAL DISTRIBUTIONS AND THEIR CHARACTERIZATIONS

Let us investigate the multivariate variants of the exponential distribution that have been studied by many authors: Gumbel, Johnson and Kotz, Marshall and Olkin and others. The main problem arising from the construction of one or another model of the multivariate exponential distribution lies in the fact that these models must possess certain characterization properties of the exponential distribution. In order to reduce the computations we consider the two dimensional case in detail and only formulate the results for higher dimensions.

Consider the two dimensional vector (ξ, η). We say that the lack of memory property holds for the vector (ξ, η) if

$$P\{\xi \geq s_1 + t_1, \eta \geq s_2 + t_2 | \xi \geq s_1, \eta \geq s_2\} = P\{\xi \geq t_1, \eta \geq t_2\}. \tag{9.1}$$

Theorem 9.1. The vector (ξ, η) satisfies relation (9.1) for all positive s_1, s_2, t_1, t_2 if and only if

$$P\{\xi \geq s, \eta \geq t\} = \exp\{-(\lambda_1 s + \lambda_2 t)\} \text{ for some } \lambda_1 > 0, \lambda_2 > 0.$$

Proof. Let $\bar{F}(s, t) = P\{\xi \geq s, \eta \geq t\}$. Setting $s_2 = t_2 = 0$ in (9.1) we obtain

$$\bar{F}_1(s_1 + t_1) \equiv \bar{F}(s_1 + t_1, 0) = \bar{F}_1(s_1)\bar{F}_1(t_1)$$

for all $s_1, t_1 \geq 0$ which, as we know from the results in Section 1, implies that $\bar{F}_1(s) = \exp\{-\lambda_1 s\}$ for some $\lambda_1 > 0$. Analogously, $\bar{F}_2(t) = \bar{F}(0, t) = \exp\{-\lambda_2 t\}$ for some $\lambda_2 > 0$.

Setting $s_2 = t_1 = 0$ in (9.1) we have

$$\bar{F}(s_1, t_2) = \exp\{-(\lambda_1 s_1 + \lambda_2 t_2)\},$$

which we also needed to prove.‖

Theorem 9.2. Let $(\xi_1, \ldots, \xi_n)$ be a random vector. The relation

$$P\{\xi_1 \geq s_1 + t_1, \ldots, \xi_n \geq s_n + t_n\} = P\{\xi_1 \geq s_1, \ldots, \xi_n \geq s_n\} P\{\xi_1 \geq t_1, \ldots, \xi_n \geq t_n\} \quad (9.2)$$

holds for all positive $s_1, t_1, \ldots, s_n, t_n$ if and only if

$$P\{\xi_1 \geq s_1, \ldots, \xi_n \geq s_n\} = \exp\{-(\lambda_1 s_1 + \cdots + \lambda_n s_n)\}$$

for some $\lambda_1 > 0, \ldots, \lambda_n > 0$.

The models of the multivariate exponential distribution obtained in Theorems 9.1-9.2 are trivial. This is the case when the components of the random vector are independent and each component has an exponential distribution.

A more interesting generalization was obtained by Marshall and Olkin.

Theorem 9.3. If the marginal distributions of a random vector (ξ, η) have an exponential distribution then (9.1) is satisfied for all positives $s_1, s_2, t_1 = t_2$ if and only if

$$\bar{F}(s, t) = \exp\{-\lambda_1 s - \lambda_2 t - \lambda_{12} \max(s, t)\}$$

for some non-negative $\lambda_1, \lambda_2, \lambda_{12}$.

Proof. Setting $s_1 = s_2$ in (9.1) we get

$$\bar{F}(s + t, s + t) = \bar{F}(s, s)\bar{F}(t, t),$$

87

which implies that $\bar{F}(s,s) = -\exp\{-\lambda s\}$ for some $\lambda > 0$. Therefore

$$\bar{F}(s+t,t) = \bar{F}(s,0)e^{-\lambda t}.$$

Furthermore, since the marginal distributions of ξ and η are exponential, we have $\bar{F}(s+t,t) = \exp\{-(\delta_1 s + \lambda t)\}$ for some $\delta_1 \geq 0$ and, since $t \geq 0$,

$$\bar{F}(x,y) = \exp\{-\lambda y - \delta_1(x-y)\} \quad \text{for} \ \ x \geq y. \tag{9.3}$$

In exactly the same way we obtain

$$\bar{F}(x,y) = \exp\{-\lambda x - \delta_2(y-x)\} \quad \text{for} \ \ x \leq y \tag{9.4}$$

for some $\delta_2 \geq 0$.

The relations (9.3) and (9.4) define $\bar{F}(x,y)$ for all $x,y \geq 0$. Since $\bar{F}(x,y)$ decreases monotonely with respect to y, then $\lambda > \delta_1$ and therefore $\lambda_2 = \lambda - \delta_1$ is non-negative. In the same way we have that $\lambda_1 = \lambda - \delta_2$ is non-negative. Let $\lambda_{12} = \delta_1 + \delta_2 - \lambda$. We now show that $\delta_1 + \delta_2 \geq \lambda$. Using the bivariate distribution given in (9.3) and (9.4), one can construct a one-dimensional distribution function

$$G(x) = 1 + \bar{F}(x,x) - \bar{F}_1(x) - \bar{F}_2(x),$$

where $\bar{F}_1(x) = \exp(-\delta_1 x)$, $\bar{F}_2(x) = \exp(-\delta_2 x)$, and $\bar{F}(x,x)$ is given by (9.3) and (9.4) with $y = x$. Therefore

$$G'(x) = \delta_2 e^{-\delta_1 x} + \delta_2 e^{-\delta_2 x} - \lambda e^{-\lambda x} \geq 0$$

for all x. Letting $x \to 0$ we find that $\delta_1 + \delta_2 - \lambda \geq 0$, from which it follows that λ_{12} is non-negative. Now substituting $\lambda = \lambda_1 + \lambda_2 + \lambda_{12}$, $\delta_1 = \lambda_1 + \lambda_{12}$, $\delta_2 = \lambda_2 + \lambda_{12}$ in (9.3) and (9.4) we obtain the assertion of the theorem.$\parallel$

If a multivariate distribution is given by the formula

$$P\{\xi_1 \geq s_1, \ldots, \xi_n \geq s_n\} = \exp\{-\sum_{i=1}^{n} \lambda_i s_i - \sum_{i<j} \lambda_{ij} \max(s_i, s_j) - \cdots - \lambda_{12\cdots n} \max(s_1, s_2, \ldots, s_n)\},$$

(9.5)

$$\lambda_{i_1,\ldots,i_k} \geq 0, \ k = 1, \ldots, n, \ 1 \leq i_1 < i_2 < \cdots < i_k \leq n,$$

then we shall call it a distribution of the Marshall-Olkin type, or, simply an M-O type distribution.

Theorem 9.4. Let $(\xi_1, \xi_2, \ldots, \xi_n)$ be a random vector with all marginal distributions of the M-O type. Relation (9.2) is satisfied for all $s_1, s_2, \ldots, s_n \geq 0$ and $t_1 = t_2 = \cdots = t_n \geq 0$ if and only if $(\xi_1, \xi_2, \ldots, \xi_n)$ has the form (9.5).

Let us introduce several additional properties of M-O type distributions.

Theorem 9.5. (ξ_1, ξ_2) has a bivariate M-O type distribution if and only if there exist random variables η_1, η_2 and η_{12} that have exponential distributions and such that $\xi_1 = \min(\eta_1, \eta_{12})$ and $\xi_2 = \min(\eta_2, \eta_{12})$.

Proof. Let

$$P\{\eta_1 \geq t\} = e^{-\lambda_1 t}, P\{\eta_2 \geq t\} = e^{-\lambda_2 t}, P\{\eta_{12} \geq t\} = e^{-\lambda_{12} t}.$$

If $\xi_i = \min(\eta_i, \eta_{12})$, $i = 1, 2$, then

$$P\{\xi_1 \geq t_1, \xi_2 \geq t_2\} = P\{\eta_1 \geq t_1\}P\{\eta_2 \geq t_2\}P\{\eta_{12} \geq \max(t_1, t_2)\}$$

$$= \exp\{-\lambda_1 t_1 - \lambda_2 t_2 - \lambda_{12} \max(t_1, t_2)\}.$$

The converse assertion is obvious.$\|$

It should be mentioned that only the property "$\min(\xi_1, \xi_2)$ has an exponential distribution" is not a characterization of an M-O distribution.

89

Example 9.1. Let (ξ_1, ξ_2) be a random vector such that

$$P\{\xi_1 \geq t_1, \xi_2 \geq t_2\} = \exp\{t_1 - t_2 + \frac{2t_1 t_2}{t_1 + t_2}\}, \ t_i \geq 0, \ i = 1, 2.$$

In this case, however, $P\{\min(\xi_1, \xi_2) \geq t\} = e^{-t}, \ t \geq 0$.

Theorem 9.6. The random vector (ξ_1, ξ_2) has an M-O type distribution if and only if:

(a) (ξ_1, ξ_2) has exponential marginal distributions;

(b) $\eta = \min(\xi_1, \xi_2)$ has an exponential distribution;

(c) η does not depend on $\varsigma = \xi_1 - \xi_2$.

Proof. To prove the theorem it is sufficient to prove that the conditions formulated are equivalent to the following: for a non-negative, two-dimensional random vector (ξ_1, ξ_2) with absolutely continuous marginal distributions and η and ς as defined above, the lack of memory property holds if and only if

(i) η and ς are independent,

(ii) η has an exponential distribution with mean δ^{-1} for some δ.

The necessity of these conditions follows from the direct computation of

$$P\{\eta = \min(\xi_1, \xi_2) \leq s, \ \varsigma = \xi_1 - \xi_2 \leq t\},$$

and the lack of memory property implies the existence of a $\theta > 0$ such that $P\{\eta \geq s\} = e^{-\theta s}, s \geq 0$.

Let us prove the sufficiency. From the independence of η and ς and the fact that η has an exponential distribution it follows that

$$\bar{F}(s,t) = P\{\xi_1 \geq s, \xi_2 \geq t\} = P\{\eta \geq t\} + \int_s^t P\{\varsigma \leq u - t\}dP\{\eta < u\}$$

$$= \exp\{-\theta t\} + \int_s^t P\{\varsigma \leq u - t\}\theta \exp\{-\theta u\}du$$

for $0 \leq s \leq t$. Therefore, for fixed $t > 0$ and $0 \leq s \leq t$ such that $P\{\varsigma < s - t\}$ is continuous, we have

$$\frac{\partial}{\partial s} \bar{F}(s, t) = -P\{\xi \leq s - t\}\theta \exp\{-\theta s\}.$$

Analogously, since $\bar{F}(t, t) = P\{\eta \geq t\}$, we have

$$\frac{\partial}{\partial s_1} \bar{F}(s_1 + t, s_2 + t) = \bar{F}(t, t)\frac{\partial}{\partial s_1} \bar{F}(s_1, s_2) \qquad (9.6)$$

holding for all s_1 such that $0 \leq s_1 \leq s_2, 0 \leq t$, with some fixed $s_2 > 0$ and $t > 0$. Integrating (9.6) with respect to s_1 from 0 to s_2, we find that (9.1) holds for $0 \leq s_1 \leq s_2$ and $0 < t_1 = t_2 = t$. Analogous reasoning is followed also for the case $0 \leq s_2 \leq s_1, 0 < t_1 = t_2 = t$. $\|$

The next two theorems on stability of the characterization of the multivariate exponential distribution and the M-O type distribution were obtained by Dimitrov, Klebanov and Rachev (1982).

Theorem 9.7. Let

$$\sup\{|\bar{F}(s_1 + s_2, t_1 + t_2) - \bar{F}(s_1, t_1)\bar{F}(s_2, t_2)| \text{ for } s_1, s_2, t_1, t_2 \geq 0\} = \epsilon.$$

Then

$$\frac{\epsilon}{3} \leq \inf_{\lambda_1, \lambda_2 \geq 0} \sup_{s_1, s_2 \geq 0} \{|\bar{F}(s_1, s_2) - \exp\{-\lambda_1 s_1 - \lambda_2 s_2\}|\} \leq 41\epsilon.$$

Theorem 9.8. Assume that

$$\sup\{|\bar{F}(s_1 + t, s_2 + t) - \bar{F}(s_1, s_2)\bar{F}(t, t)| \text{ for } s_1, s_2, t \geq 0\} = \epsilon \qquad (9.7)$$

and the conditions

$$\inf_{\lambda_1 > 0} \sup_{s_1 \geq 0} |\bar{F}(s_1, 0) - \exp\{-\lambda_1 s_1\}| < \delta_1,$$

$$\inf_{\lambda_2 > 0} \sup_{s_2 \geq 0} |\bar{F}(0, s_2) - \exp\{-\lambda_2 s_2\}| < \delta_2$$

are satisfied. Then there exists numbers $\lambda_1 \geq 0, \lambda_2 \geq 0$ and $\lambda_{12} \geq -\min(\lambda_1, \lambda_2)$ such that

$$\sup_{s_1, s_2 \geq 0} |\bar{F}(s_1, s_2) - \exp\{-\lambda_1 s_1 - \lambda_2 s_2 - \lambda_{12} \max(s_1, s_2)\}| \leq 21\epsilon + \max(\delta_1, \delta_2).$$

Multivariate exponential distributions can be introduced and characterized using multivariate analogues of the failure rate function. However, one can define the IFR- and DFR-distributions by different methods in just the same way that the lack of memory property was introduced by different methods.

Condition I-1. A random vector $\xi = (\xi_1, \xi_2, \ldots, \xi_n)$ has an increasing failure rate function, that is, $\bar{F}(s_1 + t_1, \ldots, s_n + t_n)/\bar{F}(t_1, \ldots, t_n)$ does not increase with $t = (t_1, \ldots, t_n)$ for all $s_1, \ldots, s_n$).

Condition D-1. A random vector $\xi = (\xi_1, \ldots, \xi_n)$ has a decreasing failure rate function, that is, $\bar{F}(s_1 + t_1, \ldots, s_n + t_n)/\bar{F}(t_1, \ldots, t_n)$ does not decrease with $t = (t_1, \ldots, t_n)$ for all $(s_1, \ldots, s_n)$.

Theorem 9.9. The random vector $(\xi_1, \xi_2, \ldots, \xi_n)$ satisfies Conditions I-1 and D-1 simultaneously if and only if ξ has a multivariate exponential distribution with independent components.

The proof follows that of Theorem 9.1.

Let us use the notation: $\tilde{e} = (1, \ldots, 1)$ and $\tilde{s} = (s_1, \ldots, s_n)$.

Condition I-2. For the random vector $\xi = (\xi_1, \ldots, \xi_n)$ and for any of its marginal distributions and for any vector $\tilde{s} \geq 0, (\bar{F}(\delta \tilde{e}) > 0)$, the ratio $\bar{F}(\tilde{s} + \delta \tilde{e})/\bar{F}(\delta \tilde{e})$ is non-increasing in δ.

Condition D-2. This is obtained from Condition I-2 by substituting "non-decreasing" for "non-increasing".

Theorem 9.10. Conditions I-2 and D-2 are satisfied simultaneously if and only if ξ has an M-O type distribution.

The proof is the same as that of Theorem 9.3.

Theorem 9.11. If the ratio $\bar{F}(\tilde{s}+\delta\tilde{e})/\bar{F}(\tilde{s})$ does not depend on $\tilde{s}$ and if when $\bar{F}(\cdot)$ is replaced by any marginal distribution the resulting ratio also does not depend on $\tilde{s}$ for any $\delta \geq 0(\bar{F}(\tilde{s}) > 0)$, then $\bar{F}(\cdot)$ is an M-O type distribution.

Proof. If $\xi = (\xi_1, \ldots, \xi_n)$ has an M-O type distribution then

$$\frac{\bar{F}(\tilde{s} + \delta\tilde{e})}{\bar{F}(\tilde{s})} = \exp\{-\delta(\sum_{i=1}^{n} \lambda_i + \sum_{i<j} \lambda_{ij} + \cdots + \lambda_{12\cdots n})\}$$

does not depend on $\tilde{s}$ for every δ.

Conversely, assume that ξ satisfies the conditions of the theorem. Then

$$\frac{\bar{F}(\tilde{s} + \delta\tilde{e})}{\bar{F}(\tilde{s})} = c(\delta),$$

where the constant $c(\delta)$ depends only on δ. Setting $\tilde{s} = \tilde{0}$ we get

$$c(\delta) = \bar{F}(\delta\tilde{e}).$$

Therefore

$$\bar{F}(\tilde{s} + \delta\tilde{e}) = \bar{F}(\tilde{s})\bar{F}(\delta\tilde{e})$$

for any $\delta > 0$. The assertion of the theorem follows from Theorem 9.4.‖

The following three characterization theorems were proved by A. Obretenov and S. Rachev using Conditions I-1 and D-1.

Theorem 9.12. Let the random vector (ξ, η) satisfy Condition I-1 or Condition D-1. If for any $s \geq 0$ and $t \geq 0$ the function $\bar{F}(s, t)$ satisfies the equation

$$\int_0^\infty \int_0^\infty \bar{F}(s + t, s_1 + t_1)\mu(dt, dt_1) = \bar{F}(s, s_1)$$

where μ is Borel measure on R_+^2 such that

$$G_\mu(u, u_1) = \int_u^\infty \int_{u_1} \bar{F}(t, t_1)\mu(dt, dt_1) > 0 \text{ for } u > 0, u_1 > 0,$$

then

$$\bar{F}(s, t) = \exp\{-\lambda_1 s - \lambda_2 t\}, \quad s \geq 0, \ t \geq 0$$

for some $\lambda_1 > 0, \lambda_2 > 0$.

Theorem 9.13. Assume that the random vector (ξ, η) with non-negative components satisfies the relation

$$E[(\xi - s_1)^{k_1}(\eta - s_2)^{k_2} \mid \xi \geq s_1, \eta \geq s_2] = \frac{k_1! k_2!}{\lambda_1^{k_1} \lambda_2^{k_2}}, \quad s_1, s_2 \geq 0$$

for some $k_1, k_2 = 1, 2, \ldots$ and $\lambda_1 > 0, \lambda_2 > 0$. If

$$\left. \frac{\partial^{\ell_1 + \ell_2} \bar{F}(s_1, s_2)}{\partial s_1^{\ell_1} \partial s_2^{\ell_2}} \right|_{s_i = 0} = (-\lambda_1)^{\ell_1}(-\lambda_2)^{\ell_2} e^{-\lambda_i s_i}$$

for $\ell_i = 0, 1, \ldots, k_i - 1, \lambda_i > 0, i = 1, 2$, then

$$\bar{F}(s_1, s_2) = \exp\{-\lambda_1 s_1 - \lambda_2 s_2\}, \quad s_1 \geq 0, \ s_2 \geq 0.$$

Theorem 9.14. Let

$$\int_0^\infty \bar{F}(s_1 + t, s_2 + t)\mu(dt) = \bar{F}(s_1, s_2), \quad s_1, s_2 \geq 0$$

where $\mu(\cdot)$ is Borel measure on $[0, \infty)$ with infinite support. If $\bar{F}(s_1, 0) = \exp\{-\theta_1 s_1\}$, $\bar{F}(0, s_2) = \exp\{-\theta_2 s_2\}$ for some $\theta_1, \theta_2 > 0$, then there exist positive λ_1, λ_2 and λ_{12} such that

$$\bar{F}(s_1, s_2) = \exp\{-\lambda_1 s_1 - \lambda_2 s_2 - \lambda_{12} \max(s_1, s_2)\}.$$

If a multivariate distribution has density $f(s_1, \ldots, s_n)$, then we can define the failure rate

94

density:

$$r(s_1, \ldots, s_n) = \frac{f(s_1, \ldots, s_n)}{\bar{F}(s_1, \ldots, s_n)}. \tag{9.8}$$

Theorem 9.15. If for a two dimensional random vector ξ with (finite Laplace transform and) exponential marginal distributions the function $r(s_1, s_2)$ is constant in s_1 and s_2, then ξ has an exponential distribution with independent components.

Proof. By assumption

$$r(s_1, s_2) = \frac{f(s_1, s_2)}{\bar{F}(s_1, s_2)} = \lambda \tag{9.9}$$

with

$$\bar{F}(s_1, 0) = \exp\{-\lambda_1 s_1\}; \; \bar{F}(0, s_2) = \exp\{-\lambda_2 s_2\}, \bar{F}(0, 0) = 1. \tag{9.10}$$

Equation (9.9) is equivalent to a second order equation in partial derivatives

$$\frac{\partial^2 \bar{F}(s_1, s_2)}{\partial s_1 \partial s_2} - \lambda \bar{F}(s_1, s_2) = 0 \tag{9.11}$$

with boundary conditions (9.10).

Let us look at the Laplace transform of $\bar{F}(s_1, s_2)$:

$$\varphi(x, y) = \int_0^\infty \int_0^\infty e^{-(xs_1 + ys_2)} \bar{F}(s_1, s_2) ds_1 ds_2.$$

Integrating by parts with respect to s_1 and using the condition $\bar{F}(0, s_2) = \exp\{-\lambda_2 s_2\}$, $s_2 \geq 0$, we get

$$\varphi(x, y) = \frac{1}{x} \left\{ \frac{1}{y^2 + \lambda^2} + \int_0^\infty \int_0^\infty e^{-(xs_1 + ys_2)} \frac{\partial \bar{F}(s_1, s_2)}{\partial s_1} ds_1 ds_2 \right\}.$$

Furthermore, integrating by parts with respect to s_2 and using (9.11) and the boundary condition $\bar{F}(s_1, 0) = \exp\{-\lambda_1, s_1\}$, $s_1 \geq 0$ we have

$$\varphi(x, y) = \frac{1}{x(y + \lambda_2)} - \frac{\lambda_1}{xy(x + \lambda_1)} + \frac{\lambda}{xy} \varphi(x, y),$$

95

which leads to

$$\varphi(x, y) = \frac{xy - \lambda_1\lambda_2}{(y + \lambda_2)(x + \lambda_1)(xy - \lambda)}, \tag{9.12}$$

and if we denote the Laplace transform of $f(x_1, s_2) = \lambda \bar{F}(s_1, s_2)$ by $\Psi(x, y)$ we have

$$\Psi(x, y) = \frac{\lambda(xy - \lambda_1\lambda_2)}{(x + \lambda_1)(y + \lambda_2)(xy - \lambda)}.$$

However, from the right-hand side of (9.12) it follows that for the Laplace transform $\varphi(x)$ to exist for all $x \geq 0, y \geq 0$, we must have $\lambda = \lambda_1\lambda_2$. Therefore

$$\Psi(x, y) = \frac{\lambda_1\lambda_2}{(\lambda_1 + x)(\lambda_2 + y)}$$

which is the Laplace transform of the product of two independent exponentially distributed random variables.‖

If the failure rate density is equal to a constant then, as the following theorem demonstrates, the distribution would not even have to be of the M-O type.

Theorem 9.16. For a given $\lambda > 0$, absolutely continuous distributions that satisfy $r(s_1, \ldots, s_n) = \lambda$ are a mixture of exponential distributions which have the form:

$$f(s_1, \ldots, x_n) = \lambda \int_0^\infty \cdots \int_0^\infty \exp\{-\sum_{i=1}^n \lambda_i s_i\} D(d\lambda_1, \ldots, d\lambda_n), \ s_i \geq 0,$$

where the probability measure D is concentrated on the set

$$\{\prod_{i=1}^n \lambda_i = \lambda, \ \lambda_i > 0, \ i = 1, \ldots, n\}.$$

Let us go on to further generalizations of the definition of the failure rate function.

Definition 9.1. Let $\xi = (\xi_1, \ldots, \xi_n)$ be a random vector with distribution function $F_\xi(\tilde{s}) = P\{\bigcap_{i=1}^n [\xi_i < s_i]\}$ and $\bar{F}_\xi(\tilde{s}) = P\{\bigcap_{i=1}^n [\xi_i \geq s_i]\}$.

96

Let us define a failure rate vector-function:

$$r_\xi(\tilde{s}) = (\frac{\partial}{\partial s_1}, \ldots, \frac{\partial}{\partial s_n})\{-\log \bar{F}_\xi(\tilde{s})\}$$

$$= (-\frac{\partial}{\partial s_1}\log \bar{F}_\xi(\tilde{s}), \ldots, -\frac{\partial}{\partial s_n}\log \bar{F}_\xi(\tilde{s})) \equiv (r_\xi(\tilde{s}_1), \ldots, r_\xi(\tilde{s}_n)).$$

Definition 9.2. If for all values of $\tilde{s} = (s_1, \ldots, s_n)$ all components of $r_\xi(\tilde{s})$ are increasing (decreasing) functions of the corresponding variables, that is, $r_\xi(\tilde{s})$ is an increasing (decreasing) function of s_j for $j = 1, 2, \ldots, n$, then we say that the distribution is a VIFR-(VDFR)-distribution.

Theorem 9.17. A multivariate distribution for which $r_\xi(\tilde{s})$ is a constant vector is an exponential distribution with independent components.

Proof. $r_\xi(\tilde{s}) = \tilde{c}$ implies

$$\frac{\partial \log \bar{F}_\xi(\tilde{s})}{\partial s_i} = -c_j \quad (j = 1, 2, \ldots, n).$$

Therefore

$$\bar{F}(\tilde{s}) = e^{-c_j s_j} g_j(s_1, \ldots, s_{j-1}, s_{j+1}, \ldots, s_n), \quad j = 1, \ldots, n.$$

Thus, $\bar{F}_\xi(\tilde{s})$ is proportional to $\exp\{-\sum_{j=1}^n c_j s_j\}$. Taking into account the boundary conditions we obtain the assertion of the theorem.‖

M-O type distributions are not absolutely continuous and can have a singular component $\bar{F}_s(\tilde{x})$. Let us look at the bivariate case.

Theorem 9.18. If $\bar{F}(x, y)$ is the tail of an M-O type distribution then

$$\bar{F}(x, y) = \frac{\lambda_1 + \lambda_2}{\delta}\bar{F}_a(x, y) + \frac{\lambda_{12}}{\delta}\bar{F}_s(x, y), \tag{9.13}$$

where $\bar{F}_s(x, y) = \exp\{-\delta \max(x, y)\}$ and $\delta = \lambda_1 + \lambda_2 + \lambda_{12}$. The absolutely continuous component is

$$\bar{F}_a(x, y) = \frac{\delta}{\lambda_1 + \lambda_2} \exp\{-\lambda_1 x - \lambda_2 y - \lambda_{12} \max(x, y)\}$$
$$- \frac{\lambda_{12}}{\lambda_1 + \lambda_2} \exp\{-\delta \max(x, y)\}.$$

Proof. From the decomposition $\bar{F}(x, y) = \alpha \bar{F}_a(x, y) + (1 - \alpha) \bar{F}_s(x, y)$ and from the form of an M-O type distribution follows:

$$\frac{\partial^2 \bar{F}(x, y)}{\partial x \, \partial y} = \alpha f(x, y) = \begin{cases} \lambda_2(\lambda_1 + \lambda_{12}) \bar{F}(x, y), & \text{if } x > y, \\ \lambda_1(\lambda_2 + \lambda_{12}) \bar{F}(x, y), & \text{if } x < y, \end{cases}$$

where $f_a(x, y)$ is the corresponding bivariate density. Integrating $\alpha f_a(x, y)$ we obtain the values of the constants in equation (9.13). ||

Theorem 9.19. If the marginal distributions have densities, then an M-O type distribution $\bar{F}(x, y)$ is absolutely continuous if

$$\delta = \lambda_1 + \lambda_2 + \lambda_{12} = f_1(0) + f_2(0), \tag{9.14}$$

where the $f_j(\cdot), j = 1, 2$ are the corresponding marginal densities.

Proof. Let

$$\bar{F}(x, y) = \alpha \bar{F}_a(x, y) + (1 - \alpha) \bar{F}_s(x, y), \ \ 0 \le \alpha \le 1,$$

where $\bar{F}_a(\cdot)$ and $\bar{F}_s(\cdot)$ are defined above. Taking the second derivative we obtain

$$\frac{\partial^2 \bar{F}(x, y)}{\partial x \, \partial y} = \alpha \bar{F}_a(x, y).$$

In particular

$$\int_{x \ge y} \alpha f_a(x, y) dx \, dy = 1 - \frac{1}{\delta} f_1(0),$$
$$\int_{x \le y} \alpha f_a(x, y) dx \, dy = 1 - \frac{1}{\delta} f_2(0).$$

98

Since $\bar{F}_a(\infty, \infty) = 0$, we have

$$\alpha = \int_0^\infty \int_0^\infty \alpha f_a(x, y) dx\, dy = 2 - \frac{1}{\delta}[f_1(0) + f_2(0)].$$

Therefore $\bar{F}(\cdot)$ is absolutely continuous and $\alpha = 1$ if (9.14) is satisfied.$\|$

APPENDIX

The theorems that will be proved below played an important role in proving the characterization of the exponential distribution by the lack of memory property and properties of independence of functions of order statistics as well as in the bounds in problems of stability.

Theorem II.1. Let $G(x)$ be a distribution function on $(0, \infty)$ that satisfies

$$1 < \int_0^\infty \exp\{\delta x\} dG(x) < \infty \qquad (II.1)$$

for some $\delta > 0$, and let Δ_G be the set of points of increase of $G(x)$. Then for any non-negative, measurable solution $H(x)$ of the integral equation

$$H(x) = \int_0^\infty H(x + y) dG(y) \qquad (II.2)$$

which also satisfies

$$\int_{x_0}^x H(u) du < \infty \text{ for all } x \geq x_0, \qquad (II.3)$$

we have

$$H(x + \omega) = H(x) \quad \text{a.s.} \qquad (II.4)$$

for every $\omega \in \Delta_0$.

Proof. Let us introduce

$$\bar{H}(x) = \exp\left(\frac{\delta x}{2}\right) \int_x^\infty e^{-\delta u/2} H(u) du$$

and show that $\bar{H}(x)$ is defined correctly. We shall prove that

$$\int_{x_0}^\infty e^{-\lambda u} H(u) du < \infty \text{ for all } \lambda > 0 \qquad (II.5)$$

follows from $(II.2)$ and $(II.3)$.

Since $\int_0^\infty dG(u) = 1$ there exists $k = k(\lambda)$ for every $\lambda > 0$ such that

$$\int_0^k e^{\lambda y} dG(y) \geq 1. \qquad (II.6)$$

From $(II.2)$ and Fubini's theorem we obtain

$$\int_{x_0}^x e^{-\lambda u} H(u) du \geq \int_0^k \int_{x_0}^x e^{-\lambda u} H(u + y) \, du \, dG(y)$$
$$= \int_0^k e^{\lambda y} \int_{x_0+y}^{x+y} e^{-\lambda u} H(u) \, du \, dG(y)$$

which leads to

$$\int_0^k e^{\lambda y} \int_{x_0}^{x+y} e^{-\lambda u} H(u) \, du \, dG(y) - \int_{x_0}^x e^{-\lambda u} H(u) du$$
$$\leq \int_0^k e^{\lambda y} \int_{x_0}^{x_0+y} e^{-\lambda u} H(u) \, du \, dG(y) = A_0 < \infty,$$

by virtue of $(II.3)$. Together with $(II.6)$ we have

$$\int_0^k e^{\lambda y} \int_x^{x+y} e^{-\lambda u} H(u) \, du \, dG(y) \leq A_0,$$

and if $\eta > 0$ is such that $G(k) - G(\eta) > 0$ we get

$$\int_x^{x+\eta} e^{-\lambda u} H(u) \, du < A_1 < \infty$$

for all $x \geq x_0$. Hence

$$\int_{x_0+n\eta}^{x_0+(n+1)\eta} e^{-2\lambda u} H(u) du \leq A_2 \exp[-\lambda \eta n]$$

for all $n \geq 0$, and therefore

$$\int_{x_0}^\infty e^{-2\lambda u} H(u) du = \sum_{n=0}^\infty \int_{x_0+n\eta}^{x_0+(n+1)\eta} e^{-2\lambda u} H(u) du < A_2 \sum_{n=0}^\infty e^{-\lambda \eta n} < \infty.$$

Since λ is arbitrary we obtain $(II.5)$.

101

Consequently $\bar{H}(x)$ is defined correctly and it is easy to show that $(II.2)$ is satisfied for $\bar{H}(x)$.

Furthermore, for all $x > x_0$ we have the following chain of inequalities

$$\int_0^\infty \int_x^{x+y} \bar{H}(u) \, du \, dG(y) \le \int_0^\infty \int_{x_0}^{x_0+y} \bar{H}(u) \, du \, dG(y)$$

$$\le \int_0^\infty e^{\delta(x_0+y)/2} \int_{x_0}^{x_0+y} e^{-\delta u/2} \bar{H}(u) \, du \, dG(y)$$

$$\le e^{-\delta x_0/2} \bar{H}(x_0) \int_0^\infty y e^{\delta(x_0+y)/2} dG(y) = A_3 < \infty$$

since $\bar{H}(x) \exp[-\delta x/2]$ does not increase and satisfies $(II.1)$. Thus

$$\int_0^k \int_x^{x+y} e^{\delta u/2} e^{-\delta u/2} \bar{H}(u) \, du \, dG(y) \le A_3$$

for all $k > 0$, and again, because $\bar{H}(x) \exp[-\delta x/2]$ is non- increasing, we have

$$\int_0^k e^{\delta x/2} e^{-\delta(x+k)/2} \bar{H}(x+k) dG(y) \le A_3$$

for all $x \ge x_0$. Choosing k so that $G(k) - G(0) > 0$ we get $\bar{H}(x+k) \le A_4 < \infty$ for all $x \ge x_0$, and once more because $\bar{H}(x) \exp[-\delta x/2]$ is non-increasing, we conclude that $\bar{H}(x)$ is bounded.

By virtue of the Choquet-Deny theorem (see [24], v. 2) we have that $\bar{H}(x+\omega) = \bar{H}(x)$ for all $\omega \in \Delta_G$. Hence,

$$e^{\delta x/2} \int_x^\infty e^{-\delta u/2} H(u) du = e^{\delta(x+\omega/2} \int_{x+\omega}^\infty e^{-\delta u/2} H(u) du = e^{\delta x/2} \int_x^\infty e^{-\delta u/2} H(u+\omega) du.$$

Thus,

$$\int_x^\infty e^{-\delta u/2} [H(u) - H(u+\omega)] du = 0 \text{ for all } x \ge x_0 \text{ and all } \omega \in \Delta_G.$$

Differentiating with respect to x, from Lebesgue's theorem we obtain the assertion of the thoerem. $\|$

102

Let us introduce two more conditions:

$$H(x + y) \leq e^{\lambda y} H(x), \quad x \geq x_0, \quad y \geq 0, \tag{II.7}$$

for some positive constant λ, and

$$m = \int_0^\infty e^{-\epsilon x} dG(x), \quad 0 < m < 1. \tag{II.8}$$

Theorem II.2. Let $H(x)$ be a non-negative, right continuous function satisfying condition $(II.7)$, and let $G(x)$ be a distribution function satisfying condition $(II.1)$. If m is given by $(II.8)$ and $H(x)$ is the solution of the functional equation

$$H(x) = \int_0^\infty H(x + y) dG(y) + R(x) H(x), \quad x \geq x_0 = 0, \tag{II.9}$$

where $R(x)$ is a real function such that

$$|R(x)| \leq R_0 e^{-\epsilon x} \leq \frac{1 - m}{4} e^{-\epsilon x}, \tag{II.10}$$

then

$$H(x) = \Delta + A(x) e^{-\epsilon x}, \quad x \geq 0, \tag{II.11}$$

where Δ is a non-negative constant and $A(x)$ is bounded:

$$|A(x)| \leq \frac{2}{1 - m} R_0 \inf |H(x)|, \quad x \geq 0. \tag{II.12}$$

We shall need a series of lemmas for the proof of this theorem.

Lemma II.2. If $H(x)$ and $G(x)$ satisfy $(II.7)$ and $(II.1)$ respectively, then the inequality

$$H(x) \geq \int_0^\infty H(x + y) dG(y) - C_1 e^{-\epsilon x} H(x), \quad x \geq x_0, \tag{II.13}$$

implies the boundedness of $H(x)$, where C_1 and ϵ are positive constants.

103

Proof. The lemma follows easily from Theorem II.1 if the support of $G(x)$ is located on the interval $[\alpha, \infty)$ where $\alpha > 0$. Indeed, making $x_1 > 0$ sufficiently large, one can make

$$A \equiv C_1(1 - e^{-\varepsilon\alpha} - C_1 e^{-\varepsilon(x_1+\alpha)})^{-1} > 0.$$

Set $H_0(x) = H(x) + Ae^{-\varepsilon x}H(x)$. Then, for $x \geq \max\{x_0, x_1\}$,

$$\int_0^\infty H_0(x+y)dG(y) = \int_0^\infty H(x+y)dG(y) + Ae^{-\varepsilon x}\int_0^\infty e^{-\varepsilon y}H(x+y)dG(y)$$

$$\leq H(x)\{1 + C_1 e^{-\varepsilon x} + Ae^{-\varepsilon(x+\alpha)}(1 + C_1 e^{-\varepsilon x})\} \leq H_0(x)$$

and the assertion follows from Theorem II.1.

For the general case let $G^{(n)}(x)$ be an n-fold convolution of $G(x)$ and let $R_n(x)$ be a real function defined by

$$H(x) = \int_0^\infty H(x+y)dG^{(n)}(y) + R_n(x)H(x). \tag{II.14}$$

Then $R_1(x)$ satisfies the inequality

$$R_1(x) \geq -C_1 e^{-\varepsilon x}, \quad x \geq x^* = \max\{0, x_0\}.$$

In addition, for any integers m and n

$$R_{m+n}(x)H(x) = R_m(x)H(x) + \int_0^\infty R_n(x+y)H(x+y)\,dG^{(m)}(y), \quad x \geq x_0, \tag{II.15}$$

and

$$R_n(x) \geq -C_1 \sum_{k=0}^{n-1}(1+C_1)^k e^{-\varepsilon x}, \quad x \geq x^*. \tag{II.16}$$

Relations $(II.15)$ and $(II.16)$ are easily proved by induction on n and m.

Now let $\varepsilon' = \min\{\lambda, \varepsilon\}$ and let d be a positive number such that $2 \geq e^{\varepsilon'd} > 3/2$. Then, by the law of large numbers we can find a natural number p for which $e^{\lambda d}G^{(p)}(d) < 1/2$ and

104

there exists a positive number A such that

$$C_2 = C_1 \sum_{k=0}^{p-1}(1+C_1)^k e^{-\epsilon' A} \leq e^{\epsilon' d} - 1 - e^{\lambda d}G^{(p)}(d) \leq 1.$$

This means that

$$m \equiv e^{(\lambda-\epsilon')d}G^{(p)}(d) = (1+C_2)e^{-\epsilon' d} < 1$$

and

$$A_1 \equiv C_1 \sum_{k=0}^{p-1}(1+C_1)^k \leq e^{\epsilon' A}.$$

We now show that

$$R_n(x) \geq -C_0 e^{-\epsilon x}, \quad x \geq x^{**} = \max\{0, x_0, A\}, \quad n = 1, 2, \ldots, \qquad (II.17)$$

where C_0 is a positive constant that may depend on $C_1, \lambda, \epsilon,$ and $G(x)$ but not on n. We prove
by induction that

$$R_{kp}(x) \geq -A_k e^{-\epsilon x}, \quad x \geq x^{**}, \quad n = 1, 2, \ldots, \qquad (II.18)$$

where

$$A_k = A_1 \sum_{i=0}^{k-1} m^i \leq e^{\epsilon' A}(1-m)^{-1}.$$

For $k = 1$ inequality $(II.18)$ follows from $(II.16)$ and the definition of A_1. Let $(II.18)$ hold
for $k \geq 1$. Then, from $(II.15)$ it follows that

$$\begin{aligned}
R_{(k+1)p}(x)H(x) &= R_p(x)H(x) + \int_0^\infty R_{kp}(x+y)H(x+y)dG^{(p)}(y) \\
&\geq -A_1 e^{-\epsilon x}H(x) - A_m e^{-\epsilon x}\int_0^\infty e^{-\epsilon y}H(x+y)\,dG^{(p)}(y).
\end{aligned}$$

But since

$$\begin{aligned}
\int_0^\infty e^{-\epsilon y}H(x+y)dG^{(p)}(y) &\leq H(x)\int_0^d e^{(\lambda-\epsilon')y}dG^{(p)}(y) + \int_d^\infty e^{-\epsilon' y}H(x+y)dG^{(p)}(y) \\
&\leq H(x)\{e^{(\lambda-\epsilon')d}G^{(p)}(d) + e^{-\epsilon' d}(1-R_p(x))\} \leq mH(x),
\end{aligned}$$

105

we have

$$R_{(k+1)p}(x) \geq -(A_1 + mA_k)e^{-\epsilon x} \geq -A_{k+1}e^{-\epsilon x}.$$

So we have proved $(II.18)$. Let n be a positive integer that is relatively prime to p. If $n < p$ then $(II.16)$ and the definition of A_1 yield

$$R_n(x) \geq -A_1 e^{-\epsilon x}, \quad x \geq x^{**}.$$

If $n > p$ then there exist positive integers k and ℓ such that

$$n = kp + \ell, \quad 1 \leq \ell < p.$$

Then it follows from $(II.15)$ that

$$R_n(x) = R_{kp+\ell}(x) = R_\ell(x) + H(x)^{-1}\int_0^\infty R_{kp}(x+y)H(x+y)dG^{(\ell)}(y)$$
$$\geq -e^{-\epsilon x}\{A_1 + A_k(1 - R_\ell(x))\} \geq -\{A_1 + A_k(1 + A_1)\}e^{-\epsilon x}.$$

Thus, for all cases $(II.17)$ is satisfied with

$$C_0 = e^{\epsilon' A}\{1 + \frac{1}{1-m}(1 + \epsilon'A)\}.$$

Now let C_3 be an arbitrary positive number larger than C_0 and let ϵ'' and η be positive numbers such that

$$C_3 \geq C_0 + \epsilon''C_3 \text{ and } e^{-\epsilon'\eta}(1 + C_0) \leq \epsilon''/2.$$

We can find a positive number n such that

$$e^{(\lambda-\epsilon')\eta}G^{(n)}(\eta) \leq \epsilon''/2.$$

For $x \geq x^{**}$ set

$$H_0(x) = H(x) + G_3 H(x)\exp(-\epsilon x)$$

106

Then

$$\int_0^\infty H_0(x+y)dG^{(n)}(y) = \int_0^\infty H(x+y)dG^{(n)}(y) + G_3 e^{-\epsilon x}\int_0^\infty e^{-\epsilon y}H(x+y)dG^{(n)}(y).$$

But since

$$0 \le \int_0^\infty H(x+y)dG^{(n)}(y) = H(x)\{1 - R_n(x)\} \le H(x)\{1 + C_0 e^{-\epsilon x}\}$$

and

$$\int_0^\infty e^{-\epsilon y}H(x+y)dG^{(n)}(y)$$

$$\le \int_0^\eta e^{-\epsilon' y}H(x+y)dG^{(n)}(y) + \int_\eta^\infty e^{-\epsilon' y}H(x+y)dG^{(n)}(y)$$

$$\le e^{(\lambda-\epsilon')\eta}G^{(n)}(\eta)H(x) + e^{-\epsilon'\eta}\{1 - R_n(x)\}H(x)$$

$$\le \epsilon'' H(x),$$

we have

$$\int_0^\infty H_0(x+y)dG^{(n)}(y) \le H(x) + C_0 e^{-\epsilon x}H(x) + \epsilon'' C_3 e^{-\epsilon x}H(x) \le H_0(x),$$

which proves the lemma.$\|$

Lemma II.2. If $H(x)$ satisfies the functional equation

$$H(x) = \int_0^\infty H(x+y)dG(y) + S_0(x),$$

then, for all $x \ge x_0$ and $u > 0$

$$|H(x+u) - H(x)| \le C_0 e^{-\epsilon x} \quad \text{for } u \in \Delta_G,$$

where

$$C_0 = \frac{2C_1}{1-m}, \quad |S_0(x)| \leq C_1 e^{-\epsilon x} \text{ for } x \geq x_0.$$

Proof. Let $r > 0$ and let $f(x)$ be the continuously differentiable density of a distribution concentrated on $[0, r]$. Set

$$\bar{H}(x) \equiv \int_0^r H(x+y)f(y)dy, \quad x \geq x_0.$$

Let $u \in \Delta_G$ and set $K(x) = \bar{H}(x+u) - \bar{H}(x)$ and

$$S(x) \equiv \int_0^r \{S_0(x+y+u) - S_0(x+y)\}f(y)dy.$$

Then there exists a positive constant C_2 which may depend on r, and $f(\cdot)$ is such that

$$|\bar{H}(x) - \bar{H}(x')| \leq C_2|x - x'|, \quad x, x' \geq x_0$$

and

$$|K(x)| \leq C_2, \quad |H(x)| \leq C_2, \quad x \geq x_0.$$

Also we have $|S(x)| \leq 2C_1 e^{\epsilon x}, x \geq x_0$. Moreover, $K(x)$ satisfies the equation

$$K(x) = \int_0^\infty K(x+y)dG(y) + S(x).$$

In particular

$$|K(x)| \leq \int_0^\infty |K(x+y)|dG(y) + 2C_1 e^{-\epsilon x}.$$

Iterating, with $n = 1, 2, \ldots$ we obtain

$$|K(x)| \leq \int_0^\infty |K(x+y)|dG^{(n)}(y) + 2C_1(1 + m + \ldots + m^{n-1})e^{-\epsilon x}.$$

From the law of large numbers it follows that

$$\lim_{n \to \infty} G^{(n)}(\sqrt{n}) = 0.$$

108

Furthermore, let $a = \overline{lim}_{x \to \infty} K(x)$. Then

$$|K(x)| \le a + C_0 e^{-\varepsilon x}, \ x \ge x_0 \ \text{and} \ a = 0.$$

For any $\varepsilon_1 > 0$ we can find an $x_1 > x_0$ such that $|K(x)| \le a + \varepsilon_1$ for all $x > x_1$. Now let $x > x_0$ and n be such that $\dot{x} + \sqrt{n} \ge x_1$. Then

$$|K(x)| \le \int_0^\infty |K(x+y)| dG^{(n)}(y) + C_0 e^{-\varepsilon x}$$

$$\le \int_0^{\sqrt{n}} |K(x+y)| dG^{(n)}(y) + \int_{\sqrt{n}}^\infty |K(x+y)| dG^{(n)}(y) + C_0 e^{-\varepsilon x}$$

$$\le C_2 G^{(n)}(\sqrt{n}) + (a + \varepsilon_1)(1 - G^{(n)}(\sqrt{n})) + C_0 e^{-\varepsilon x},$$

as $n \to \infty : |K(x)| \le a + \varepsilon_1 + C_0 e^{-\varepsilon x}$ or, in as much as ε_1 is arbitrary, $|K(x)| \le a + C_0 e^{-\varepsilon x}$.

Furthermore, without loss of generality, we have

$$a = \overline{lim} K(x) \ge -\underline{lim} K(x).$$

If $a > 0$ then one can find positive ε_1 and δ and integer L: $a > 3\varepsilon_1 > 0, \varepsilon_1 > C_2 \delta > 0, u > \delta > 0$ and $L(a - 3\varepsilon_1) > 3C_2$. Let $A = [u - \delta, u + \delta]$ and set

$$\eta = \int_A dG(x), \quad u \in \Delta_G.$$

Then, for $x \geq x_0$

$$K(x) = \int_0^\infty K(x+y)dG(y) + S(x)$$

$$\leq \int_A K(x+y)dG(y) + \int_{\bar{A}} |K(x+y)|dG(y)$$

$$\leq \eta \sup_{y \in A} K(x+y) + (1-\eta)(a + C_0 e^{-\epsilon x}) + 2C_1 e^{-\epsilon x}$$

or

$$K(x) \leq \eta K(x+u_1) + a(1-\eta) + B e^{-\epsilon x},$$

where $u_1 \in A$ and $B = (1-\eta)C_0 + 2C_1$. In just the same way we get

$$K(x+u_1) \leq \eta K(x+u_1+u_2) + a(1-\eta) + B e^{-\epsilon(x+u_1)}$$

for $u_2 \in A$. Combining the last two inequalities we arrive at

$$K(x) \leq \eta^2 K(x+u_1+u_2) + a(1-\eta^2) + B(1+\eta)e^{-\epsilon x}.$$

Integrating, we get

$$K(x) \leq \eta^2 K(x+u_1+\ldots+u_k) + a(1-\eta^k) + B(1+\eta+\ldots+\eta^{k-1})e^{-\epsilon x}$$

for $x \geq x_0$ and $k = 1, 2, \ldots$, where all $u_i \in A$. Let us choose x^* so that

$$B e^{-\epsilon x^*} \sum_{k=1}^L (\eta^{-1} + \eta^{-2} + \ldots + \eta^{-k}) \leq C_2.$$

From the definition of a we find $x_1 (\geq x^*)$:

$$a - \epsilon_1 \eta^L \leq K(x_1).$$

110

This leads to

$$a - \varepsilon_1 \eta^{L-k} \leq K(x_1 + u_1 + \ldots + u_k) + Be^{-\varepsilon x_1}(\eta^{-1} + \ldots + \eta^{-k}).$$

Combining this inequality by parts, for $k = 1, \ldots, L$ we get

$$
\begin{aligned}
L(a - \varepsilon_1) &\leq \sum_{k=1}^{L} K(x_1 + u_1 + \ldots + u_k) + C_1 \\
&= \sum_{k=2}^{L} \{\bar{H}(x_1 + u_1 + \ldots + u_{k-1} + u) - \bar{H}(x_1 + u_1 + \ldots + u_k)\} \\
&\quad + \bar{H}(x_1 + u_1 + \ldots + u_L + u) - \bar{H}(x_1 + u_1) + C_1 \\
&\leq (L-1)\delta C_2 + 3C_2 \leq L(a - 2\varepsilon_1),
\end{aligned}
$$

which is a contradiction, so $a = 0$.

Furthermore, we get

$$|K(x)| = \left| \int_0^r \{H(x + u + y) - H(x + y)\}f(y)dy \right| \leq C_0 e^{-\varepsilon x}, \quad x \geq x_0.$$

Since r is arbitrary and $H(x)$ is continuous on the right, the assertion of the lemma follows from the last equality.‖

Lemma II.3. We assume that $u \in \Delta_G$ and $u > 0$. Then there exists a periodic function $\Delta_u(x)$ such that $\Delta_u(u + x) = \Delta_u(x)$ for all x and

$$|H(x) - \Delta_u(x)| \leq C_0(1 - e^{-\varepsilon u})^{-1}e^{-\varepsilon x}, \quad x \geq x_0.$$

Proof. Let $u \in \Delta_G$. Then from Lemma II.2 it follows that

$$|H(x + \ell u) - H(x)| \leq C_0 \sum_{k=0}^{\ell-1} e^{-\varepsilon k u}e^{-\varepsilon x}, \quad x \geq x_0, \, \ell = 1, 2, \ldots \, .$$

111

In particular, $\lim_{\ell\to\infty} H(x + \ell u)$ exists for all $x \geq x_0$ and the required inequality holds with

$$\Delta_u(x) = \lim_{\ell\to\infty} H(x + \ell u). \parallel$$

Proof of Theorem II.2. If $H(x_1) = 0$ for some $x_1 > 0$ then $H(x) = 0$ for all $x \geq x_1$ from condition $(II.7)$, and therefore $H(x)$ is bounded. And if $H(x)$ is positive, the boundedness follows from Lemma II.1. We can then apply Lemma II.3 and conclude that $H(x)$ can be represented as

$$H(x) = \Delta + A(x)e^{-\epsilon z}, \quad x \geq 0.$$

Substituting this expression in $(II.9)$ we get

$$A(x)e^{-\epsilon z} = e^{-\epsilon z} \int_0^\infty A(x+y)e^{-\epsilon y}dG(y) + R(x)(\Delta + A(x)e^{-\epsilon z}), \quad x \geq 0.$$

Hence it follows that

$$|A(x)| \leq m\sup_{z\geq 0}|A(x)| + R_0\Delta + \frac{1-m}{4}\sup_{z\geq 0}|A(x)|, \quad x \geq 0,$$

or

$$\sup_{z\geq 0}|A(x)| \leq \frac{4}{3(1-m)}R_0\Delta. \qquad (II.19)$$

On the other hand we have

$$|H(x)| \geq \Delta - \sup_{z\geq 0}|A(x)| \geq (1 - \frac{4}{3(1-m)}R_0)\Delta \geq \frac{2}{3}\Delta.$$

The bound $(II.12)$ follows from the last inequality and $(II.19)$.$\parallel$

COMMENTS

Section 1.

A.Cauchy [53] and G. Darboux [60] proved Theorem 1.1; G. Young [143] gave another proof. T.A. Azlarov, M. M. Sultanova and A. A. Džamirzaev [5] obtained Theorem 1.2; earlier a weaker bound was obtained by T. A. Azlarov in [1] (in connection with this, see also [25]). N. Krishnaji [102] proved Theorem 1.3. B. Ramachandran [118] and R. Shimizu [129, 130] proved Theorem 1.5 by different methods. Here we followed the method of Shimizu. The history of this subject is provided in these papers. Theorem 1.4 is ideologically close to Theorem 1.5 and was proved by G. Marsaglia and A. Tubilla [107]. Theorem 1.6 was proved by R. Shimizu in [132]. Example 1.1 was constructed by T. A. Azlarov in his doctoral dissertation [2] and was first published in [3]. In [65] R. T. Durrett and S.G. Ghurye considered the lack of memory property in sets others than the set of positive real numbers. Problems concerned with the distribution of the remaining lifetime (Balkema and Hann [39], Kotz and Johnson [100], Shantaram and Harkness [128]) are similar to problems of the "lack of memory property" type as are problems connected with the waiting time paradox (Gupta [86, 87], Hawkins and Kotz [89]).

Section 2.

Theorems 2.1 and 2.3 appear here for the first time, apparently. Theorems 2.2 and 2.4 have very simple proofs; it seems to be more difficult to find the origin of the results than it is to prove them. Theorem 2.5 and Corollary 2.1 were proved by the authors. A weaker result was formulated by the authors in [4]. Inequality (2.19) for a more general class of distribution functions was obtained in [105]. The work of A. Obretenov [18, 112], A. D. Solovév [23], Heyde and Leslie [90] is closely related to the results of this section. Example 2.1 was constructed by the authors.

113

Section 3.

Most of the results in this section are taken from H. David's book [9]. More examples and other useful properties of order statistics including those for the exponential distribution are in the book by Sarhan and Greenberg [21]. Properties of order statistics and also their linear forms were studied by Rogers [122] and Tanis [139]. Distributions of record values were considered in Karlin's book [11] and also in [120, 134, 140]. Theorem 3.2 was proved by Sen [125]; the proof we cite was communicated to us by V. A. Egorov.

Section 4.

Theorem 4.1 was proved by M.Fisz [74]. Further generalizations and sharpening of this theorem were obtained by T. Ferguson in [71, 73]. A series of generalizations in other directions were obtained by H. J. Rossberg in [123, 124] (results from the theory of functions of a complex variable were used in his proofs) and by J. Galambos [78]. Theorem 4.2 was proved by J. Galambos in [76] (see also [79]). The work of Erickson and Guess [69] and Srivastava [135] is also related. R. Shimizu [129] proved Theorem 4.3. H. Cundy [57], Reinhardt [119] and Shanbhag [126] proved this theorem for the case $G(x) = x$, and T. A. Azlarov, M. M. Sultanova and A. A. Džamirzaev [5] proved it for the case $G(x) = x^2$. For the case $G(x) = x^k$ where k is an arbitrary natural number, the theorem was proved by T. A. Azlarov [2,3] as well as by Ju. V. Derjagin and N. A. Polesicka [8], O. M. Sahobov and A. A. Gešev [22], Laurent [103] and A. Dallas [59]. S. T. Huang [93] studied the case $G(x) = x^\alpha$ (where α is any positive number) using a completely different approach. We note that the result of L. B. Klebanov [13], obtained by using the results of I. C. Gohberg and M. G. Kreĭn [16], extends Theorem 4.3. R. Shimizu [132] proved Theorem 4.4. For the cases $G(x) = x$ and $G(x) = x^2$ the bound for the stability under weaker restrictions was obtained in [5] using the methods of [7]. A survey of results on the stability of characterizations of distributions, including the exponential, is contained

114

in the work of Lukacs [104]. Ahsanullah [31] proved Theorem 4.6 and Nagaraja [111] proved Theorem 4.7. Tata's results [140] are closely related to these two theorems. Theorem 4.8 was proved by O. M. Sakhobov and A. A. Gešev [22]. As we mentioned in Section 4, equations that may be thought of as containing conditional moments of random variables arose in the study of independence of functions of order statistics. The work of Berk [46], Bolger and Harkness [50], Hamdan [88], Beg and Kirmani [44] and Talwalker [138] is devoted to the recovery of the form of a distribution with conditional moments. Various functional equations that arise in characterization problems of this type were studied by Aczél [27].

Section 5.

Theorem 5.1 was proved by R. Shimizu [131]. Desu's result [63] and a whole series of other well known results that may be found in the book [79] followed from Corollary 5.2. Theorems 5.2-5.6 were proved in the work of Ahsanullah [28-31]; the history of the question may be traced in these papers. The article by Grosswald, Kotz and Johnson [84] (see also [142]), which in some special cases overlaps Ahsanullah's results, touches on this. Characterization problems associated with the same distribution of functions of order statistics were studied also by Arnold and Isaacson [37] and Basu [41].

Section 6.

Many authors have worked on the unique determination of a distribution function by sets of moments of order statistics: Ali [32] Arnold and Meeden [38], Chan [54], Chong [55], Huang [91, 92], Kadane [97, 98], Mallows [106], Pollak [114], and others. A history is presented in detail in [79] (Section 3.4). This section is based on the work of Hwang [94] and Govindara-julu [81-83]. Hwang's results are formulated in Theorems 6.1-6.5. They were proved using well-known theorems of Müntz-Szász and Bernstein and apparently have not been improved on. Theorem 6.6 was proved by Mkrtcjan [17]. Govindarajulu's theorems, proved for the exponential distribution, in fact illustrate the general theorems of Huang.

Section 7.

Since the geometric distribution is the discrete analogue of the exponential distribution it is natural that it is studied simultaneously with the exponential distribution. Some of the first characterizations were obtained by Ferguson [72], Crawford [56], Shanbhag [126], Srivastava [136], Puri and Rubin [115, 116]. A survey of this and other work was done by Galambos in the second part of [77] and Kotz in [99]. Theorem 7.1 is the discrete analogue of the lack of memory property and was proved in this form in [24]. Arnold [36], using the results of Shanbhag [127], proved Theorem 7.2. Theorem 7.3 was proved in the work of Emad and Govindarajulu [66] and is the analogue of Theorem 4.1. Theorems 7.4 and 7.5 were proved by the authors and sharpen A. Obretenov's results [20]. Dallas [58] proved Theorem 7.6. Among the works devoted to the geometric distribution, the work of Kirmani and Alam [101] and Berg [45] should be mentioned. Both the exponential and the geometric distributions can be characterized by the maximum entropy property in a certain class of distributions [80].

Section 8.

Results conected with Theorem 8.1 were obtained by T. A. Azlarov, M. M. Sultanova and A. A. Džamirzaev [5]. Szantai [137] and Arnold [34, 35] obtained similar results independently. Arnold considered a more general case in [35]. Riedel [121] proved Theorem 8.2; Corollary 8.1 was obtained earlier by Bosch [51]. L. B. Klebanov and I. A. Melamed [14] proved Theorem 8.3. Analogous functional equations arose in [12].

Section 9.

Many authors have studied the multivariate exponential distribution: Arnold [33], Basu and Block [42, 43] and [48, 49], Bildikar and Patil [47], Gumbel [85], Downton [64], Johnson and Kotz [95], Kabe [96], Paulson [113], Puri and Rubin [117], Freund [75], Esary and Marshall

116

[70]. The most essential results here belong to Marshall and Olkin [108, 110]. They proved Theorems 9.3, 9.4 and a series of others. A more detailed historical survey and a series of other results are given in [79, Chapter 5].

Appendix.

The idea of using the Choquet-Deny Theorem to solve convolution type equations belongs to Shanbhag and Shimizu. Shanbhag [127] did this for discrete distributions and Shimizu [129] for the general case. Brandofe and Davies [52] proved Theorem II.1. Here we have given Shimizu's simplified proof which has less rigid restrictions on the functions $G(x)$ and $H(x)$. Further generalizations and sharpening of these results have been made by Davies [61] and Ramachandran [118]. Shimizu [132] proved Theorem II.2. A series of generalizations to other distributions were proved by Shimizu and Davies [133].

REFERENCES

1. Azlarov, T. A. (1972). Stability of characterizing properties of the exponential distribution. *Litovsk. Mat. Sb.*, 12, 5–9, 199. [MR 47 (1974) 7847]

2. Azlarov, T. A. (1972). Study of the mathematical theory of queueing. Doctoral dissertation, Tashkent.

3. Azlarov, T. A. (1979). Characteristic properties of the exponential distribution and their stability. *Limit Theorems, Random Processes and Their Applications*, 230 3–14, "Fan",Tashkent. [MR 82a (1982) 62028]

4. Azlarov, T. A., and Volodin, N. A. (1981). On proximity to the exponential distribution with monotone failure intensity. *Theory of Probability and Its Applications,*, 26, 650.

5. Azlarov, T. A., Džamirzaev, A. A., and Sultanova, M. M. (1972). Characterizing properties of the exponential distribution and their stability. *Random Processes and Statistical Inference, No. II*, 94, 10–19, Izdat. "Fan" Uzbek. SSR, Tashkent. [MR 48 (1974) 3150]

6. Barlow, R., and Proschan, F. (1965). *Mathematical Theory of Reliability*, New York: Wiley.

7. Bellman, R. (1953). *Stability Theory of Differential Equations*. New York: McGraw Hill.

8. Derjagin, Ju.V., and Polesickaja, N. A. (1975). A characterization of the exponential distribution by a lack of memory type property. *Proceedings of the Moscow Institute of Electronic Machine Construction, 44th issue*, Moscow, 192–198.

8a. Dimitrov, B., Klebanov, L. and Rachev, S. (1982). Stability of a characterization of the exponential law. *Stability Problems for Stochastic Models*, Proceedings of the Sixth International Seminar, Moscow, USSR, (edited by V. V. Kalashnikev and V.M. Zolotarev).

9. David, H. A. (1981). *Order Statistics*, (2nd edition). New York: Wiley.

119

10. Kagan, A. M., Linnik, Yu. V., and Rao, C. R. (1973). *Characterization Problems in Mathematical Statistics*. New York: Wiley.

11. Karlin, S. (1968). *A First Course in Stochastic Processes*. New York: Academic Press.

12. Klebanov, L. B. (1978). Some problems of characterization of distributions arising in reliability theory. *Teor. Verojatnost. i Primenen.* 23, 828–831. [Theory of Probability and Its Applications, 23, 798–801.] [MR 81e (1981) 62010]

13. Klebanov, L. B. (1980). Some results connected with the characterization of the exponential distribution. *Teor. Verojatnost. i Primenen.*, 25, 628–633. [Theory of Probability and Its Applications, 25, 617–622.] [MR 81k (1981) 62015]

14. Klebanov, L. B., and Melamed, I. A. (1979). Stability of the characterization of the exponential distribution by the discretization property. Problems of Stability of Stochastic Models, *Proceedings of the Fourth All Union Seminar, Palanga*, 124, 63–64, [MR 84c (1984) 62025]

15. Klebanov, L. B., and Yanushkyavichene, O. L. (1982). Stability of the characterization of the exponential law. *Litovsk. Mat. Sb.*, 22, 3, 103–111. [MR 84e (1984) 62031]

16. Gohberg, I.C. and Kreĭn, M. G. (1958). Systems of integral equations on the half-line with kernels depending on the difference of the arguments. *Uspehi Mat. Nauk (N.S.)*, 13, 3–72. [MR 21 (1960) 1506]

17. Mkrtchyan, S. T. (1981). An estimate of the closeness of distributions that have coincident mean values of order statistics. *Akad. Nauk Armyan. SSR Dokl.*, 72, 195–202. [MR 83b (1983) 62088]

18. Obretenov, A. (1970). A property of the exponential distribution *Fiz.-Mat. Spis. B"lgar. Akad. Nauk.*, 13, (46), 51–53. [MR 58 (1979) 24653]

19. Obretenov, A. (1971/72). Estimation of the difference between an IFR distribution and the exponential distribution. *Annuaire Univ. Sofia Fac. Math.*, 66, 107–112. [MR 54 (1977) 4010]

20. Obretenov, A. (1978). Upper bound of the absolute difference between a discrete IFR distribution and a geometric one. *Proceedings of the Sixth Conference on Probability Theory (Brasov, 1979)*, 177–184, Ed. Acad. R.S. Romãnia, Bucharest, 1981. [MR 82m (1982) 62229]

21. Sarhan, A. E., and Greenberg, B. G. (1962). *Contributions to Order Statistics*, New York: Wiley.

22. Sakhabov, O. M., and Gešev A. A. (1974). Characterizing properties of the exponential distribution, Natura University, Plovdiv, 7, 25–28.

23. Solov'ev, A. D. (1975). *Foundations of the Mathematical Theory of Reliability*, Moscow.

24. Feller, W. (1968). *An Introduction to Probability Theory and Its Appliecations*, (3rd Edition) Vols. 1,2. New York: Wiley.

25. Hoang, Huu Nhu (1968). A bound for the stability of a certain characterization of the exponential law. *Lithuanian Mathematical Journal*, 8, 175–177.

26. Aczél, J. (1961) Vorlesungen über Funktionalgleichungen und ihre Anwendungen. Basel-Stuttgart: Birkhäuser Verlag. [MR 23A (1962) 1959]

27. Aczél, J. (1975). General solution of a functional equation connected with a characterization of statistical distributions. In G. P. Patil, S. Kotz, and J. K. Ord (Eds.), *Statistical Distributions in Scientific Work*, Vol. 3, (pp. 47–55). Dordrecht-Holland: D. Reidel Publishing Co.

28. Ahsanullah, M. (1975) A characterization of exponential distribution. In G. P. Patil, S. Kotz, and J. K. Ord (Eds.), *Statistical Distributions in Scientific Work*, Vol. 3, (pp. 131-135). Dordrecht-Holland: D. Reidel Publishing Co.

29. Ahsanullah, M. (1977). A characteristic property of the exponential distribution. *Ann. Statist.*, 5, 580–582. [MR 55 (1978) 11458]

30. Ahsanullah, M. (1978). On a characterization of the exponential distribution by spacings. *Ann. Inst. Statist. Math.*, 30, 163– 166.

31. Ahsanullah, M. (1978). Record values and the exponential distribution. *Ann. Inst. Statist. Math.*, 30, 429–433.

32. Ali M. M. (1976). An alternative proof of order statistics with moment problem. *Canad. J. Statist*, 4, 151–153. [MR 56 (1978) 6958]

33. Arnold, B. C. (1967). A note on multivariate distributions with specified marginals. *J. Amer. Statist. Assoc*, 62, 1460–1461.

34. Arnold, B. C. (1973). Some characterizations of the exponential distribution by geometric compounding. *SIAM J. Appl. Math.*, 24, 242- -244. [MR 48 (1974) 12674]

35. Arnold, B. C. (1975). A characterization of the exponential distribution by multivariate geometric compounding. *Sankhyā. Ser. A*, 37, 164–173. [MR 55 (1978) 13662]

36. Arnold, B. C. (1980). Two characterizations of the geometric distribution. *J. Appl. Probab.*, 17, 570–573. [MR 82h (1982) 62024]

37. Arnold, B. C. and Isaacson, D. (1976). On solutions to $\min(X, Y) \stackrel{d}{=} aX$ and $\min(X, Y) \stackrel{d}{=} aX \stackrel{d}{=} bY$. *Zeitschrift für Wahrscheinlichkeitstheorie und Verwandte Gebiete*, 35, 115–119. [MR 54 (1977) 3904]

38. Arnold, B. C. and Meeden, G. (1975). Characterization of distributions by sets of moments of order statistics. *Ann. Statist.*, 3, 754–758. [MR 51 (1976) 14433]

39. Balkema, A. A. and de Haan, L. (1974). Residual life time at great age. *Ann. Probab.*, 2, 792–804 [MR 50 (1975) 11504]

40. Barlow, R. E., Marshall, A. W. and Proschan, F. (1963). Properties of probability distributions with monotone hazard rate. *Ann. Math. Statist.*, 34, 375–389. [MR 30 (1965) 1559]

41. Basu, A. P. (1965). On characterizing the exponential distribution by order statistics. *Ann. Inst. Statist. Math.*, 17, 93– 96. [MR 31 (1966) 2799]

42. Basu, A. P. (1971). Bivariate failure rate. *J. Amer. Statist. Assoc.*, 66, 103–104.

43. Basu, A. P. and Block, H. W. (1975). On characterizing univariate and multivariate exponential distributions with applications. In G. P. Patil, S. Kotz, and J. K. Ord (Eds.), *Statistical Distributions in Scientific Work*, Vol. 3 (pp 399-421). Dordrecht-Holland: D. Reidel Publishing Co.

44. Beg, M. I. and Kirmani, S. N. U. A. (1974). On a characterization of exponential and related distributions. *Australian J. Statist.*, 16, 163–166 (correction in v. 18, p. 85). [MR 52 (1976) 9437, MR E55 (1978) 11459]

45. Berg S. (1978). Characterization of a class of discrete distributions by properties of their moment distributions. *Commun. Statist.*, A7, 785–789. [MR 80c(1980) 62013]

46. Berk, R. H. (1977). Characterizations via conditional distributions. *J. Appl. Probab.*, 14, 806–816. [MR 57 (1979) 10872]

47. Bildikar, S. and Patil, G. P. (1968). Multivariate exponential-type distributions. *Ann. Math. Statist.*, 39, 1316–1326. [MR 37 (1969) 7020]

48. Block, H. W. (1977). A characterization of a bivariate exponential distribution. *Ann. Statist.*, 5, 808–812. [MR 55 (1978) 13663]

49. Block, H. W. and Basu, A. P. (1974). A continuous bivariate exponential extension. *J. Amer. Statist. Assoc.*, 69, 1031–1037. [MR 55 (1978) 6685]

50. Bolger, E. M. and Harkness, W. L. (1965). Characterizations of some distributions by conditional moments. *Ann. Math. Statist.*, 36, 703–705. [MR 30 (1965) 5339]

51. Bosch, K. (1977). Eine Characterisierung der Exponentialverteilungen. *Zeitschrift für Angewandte Mathematik und Mechanik*, 57, 609–610. [MR 57 (1979) 1728]

52. Brandhofe, T. and Davies, L. (1980). On a functional equation in the theory of linear statistics. *Ann. Inst. Statist. Math.*, A, 32, 17–23. [MR 81F (1981) 45013]

53. Cauchy, A. L. (1821). Cours d'analyse de l'Ecole Polytechnique. *Analyse algébrique*, 1, Paris.

54. Chan, L. K. (1967). On a characterization of distributions by expected values of extreme order statistics. *Amer. Math. Monthly*, 74, 950–951. [MR 36 (1968) 4751]

55. Chong, K. M. (1977). On characterizations of the exponential and geometric distributions by expectations. *J. Amer. Statist. Assoc.*, 72, 160–161. [MR 55 (1978) 13628]

56. Crawford, G. B. (1966). Characterization of geometric and exponential distributions. *Ann. Math. Statist.*, 37, 1790–1795. [MR 34 (1967) 6825]

57. Cundy, H. (1966). Birds and atoms. *Math. Gazette*, 50, 294–295.

58. Dallas, A. C. (1974). A characterization of the geometric distribution. *J. Appl. Probab.*, 11, 609–611. [MR 50 (1975) 15056]

59. Dallas, A. C. (1979). On the exponential law. *Metrika*, 26, 105–108. [MR 81F (1981) 62022]

60. Darboux, G. (1875). Sur la composition des forces en statique. *Bull. Sci. Math.*, 9, 281–288.

61. Davies, L. (1981). A theorem of Deny with applications to characterization problems. *Lect. Notes Math.*, 861, 35–41. [MR 84H (1984) 62024]

62. Davis, D. J. (1952). An analysis of some failure data. *J. Amer. Statist. Assoc.*, 47, 113–150.

63. Desu, M. M. (1971). A characterization of the exponential distribution by order statistics. *Ann. Math. Statist.*, 42, 837– 838.

64. Downton, F. (1970). Bivariate exponential distributions in reliability theory. *J. Roy Statist. Soc.*, Ser. B, 32, 408–417. [MR 44 (1972) 4855]

65. Durrett, R. T. and Ghurye, S. G. (1976). Waiting times without memory. *J. Appl. Probab.*, 13, 65–75. [MR 53 (1977) 4323]

66. El-Neweihi Emad and Govindarajulu, Z. (1970). Characterizations of geometric distribution and discrete IFR (DFR) distributions using order statistics. *J. Statist. Plann. Inference*, 3, 85–90.

67. Epstein, B. and Sobel, M. (1953). Life testing. *J. Amer. Statist. Assoc.*, 48, 486–502. [MR 15 (1954) 143]

68. Epstein, B. and Sobel, M. (1954). Some theorems relevant to life testing from an exponential distribution. *Ann. Math. Statist.*, 25, 373–381. [MR 15 (1954) 810]

69. Erickson, K. B. and Guess, H. A. (1973). A characterization of the exponential law. *Ann. Probab.*, 1, 183–185. [MR 49 (1975) 11648]

70. Esary, J. D. and Marshall, A. W. (1974). Multivariate distributions with exponential minimums. *Ann. Statist.*, 2, 84–98. [MR 50 (1975) 15144]

71. Ferguson, T. S. (1964). A characterization of the exponential distribution. *Ann. Math. Statist.*, 35, 1199–1207. [MR 29 (1965) 5322]

72. Ferguson, T. S. (1965). A characterization of the geometric distribution. *Amer. Math. Monthly*, 72, 256–260. [MR 33 (1967) 1869]

73. Ferguson, T. S. (1967). On chararcterizing distributions by properties of order statistics. *Sankhyā*, A, 29, 265–278. [MR 37 (1969) 2391]

74. Fisz, M. (1958). Characterization of some probability distributions. *Skand. Aktuarietidskr.*, 41, 65–67. [MR 22 (1961) 273]

75. Freund, J. E. (1961). A bivariate extension of the exponential distribution. *J. Amer. Statist. Assoc.*, 56, 971–977. [MR 24A (1962) 2465]

76. Galambos, J. (1972). Characterization of certain populations by independence of order statistics. *J. Appl. Probab.*, 9, 224–230. [MR 45 (1973) 1214]

77. Galambos, J. (1975). Characterizations of probability distributions by properties of order statistics. I; II. In G. P. Patil, S. Kotz, and J. K. Ord (Eds.), *Statistical Distributions in Scientific Work*, Vol. 3, (pp. 71–88; 89–101). Dordrecht-Holland: D.Reidel Publishing Co.

78. Galambos, J. (1975). Characterizations in terms of properties of the smaller of two observations. *Commun.Statist.*, 4, 239–244. [MR 51 (1976) 4492]

79. Galambos, J. and S. Kotz (1978). Characterizations of probability distributions. *Lecture Notes Math.*, 675, viii+169. [MR 80A (1980) 62022]

80. Gokhale, D. V. (1975). Maximum entropy characterizations of some distributions. In G. P. Patil, S. Kotz, and J. K. Ord (Eds.), *Statistical Distributions in Scientific Work*, Vol. 3, (pp. 299–304). Dordrecht-Holland: D.Reidel Publishing Co.

81. Govindarajulu, Z. (1966). Characterization of the exponential and power distributions. *Skand. Aktuarietidskr.*, 49, 132–136. [MR 36 (1968) 6086]

82. Govindarajulu, Z. (1975). Characterization of the exponential distribution using lower moments of order statistics. In G. P. Patil, S. Kotz, and J. K. Ord (Eds.), *Statistical Distributions in Scientific Work*, Vol. 3, (pp. 117–129). Dordrecht-Holland: D.Reidel Publishing Co.

83. Govindarajulu, Z., Huang, J. S. and Saleh, A. K. Md. E. (1975). Expected value of the spacings between order statistics. In G. P. Patil, S. Kotz, and J. K. Ord (Eds.), *Statistical Distributions in Scientific Work*, Vol. 3, (pp. 143–147). Dordrecht-Holland: D.Reidel Publishing Co.

84. Grosswald, E., Kotz, S. and Johnson, N. L. (1980). Characterizations of the exponential distribution by relevation type equations. *J. Appl. Probab.*, 17, 874–877. [MR 81H (1981) 62025]

85. Gumbel, E. J. (1960). Bivariate exponential distributions. *J. Amer. Statist. Assoc.*, 55, 698–707. [MR 22 (1961) 7191]

86. Gupta, R. C. (1976). Some characterizations of distributions by properties of their forward and backward reccurrence times in a renewal process. *Scand. J. Statist.*, 3, 215–216. [MR 54 (1977) 14184]

87. Gupta, R. C. (1979). Waiting time paradox and size based sampling. *Commun. Statist.*, Ser. A, 8, 601–607. [MR 80M (1980) 62012]

88. Hamdan, M. A. (1972). On a characterization by conditional expectations. *Technometrics*, 14, 497–499.

89. Hawkins, D. and Kotz, S. (1976). A clocking property of the exponential distribution. *Australian J. Statist.*, 18, 170–172. [MR 57 (1979) 1730]

90. Heyde, C. C. and Leslie, J. R. (1976). On moment measures of departure from the normal and exponential laws. *Stochast. Process. Appl.*, 4, 317–328. [MR 54 (1977) 14039]

91. Huang, J. S. (1975). Characterization of distributions by the expected values of the order statistics. *Ann. Inst. Statist. Math.*, 27, 87–93. [MR 51 (1976) 14340]

92. Huang, J. S. and Hwang, J. S. (1975). L_1-completeness of a class of beta densities. In G. P. Patil, S. Kotz, and J. K. Ord (Eds.), *Statistical Distributions in Scientific Work*, Vol. 3, (pp. 137–141). Dordrecht-Holland: D.Reidel Publishing Co.

93. Huang, S. T. (1979). Characterizations of the exponential distributions by conditional moments. *Inst. Statist. Mimeo. Ser.*, No. 1214, 9 p.

94. Hwang, J. S. (1978). A note on Bernštein and Müntz—Szász theorems with applications to the order statistics. *Ann. Inst. Statist. Math.*, 30, 167–176. [MR 80K (1980) 62028]

95. Johnson, N. L. and Kotz, S. (1975). A vector multivariate hazard rate. *J. Multivar. Anal.*, 5, 53–66. [MR 51 (1976) 2153, erratum MR 53 (1977) 9509]

96. Kabe, D.G. (1969). On characterizing the multivariate linear exponential distribution. *Canad. Math. Bull.*, 12, 567–572. [MR 40 (1970) 4990]

97. Kadane, J. B. (1971). A moment problem for order statistics. *Ann. Math. Statist.*, 42, 745–751. [MR 45 (1973) 1328]

98. Kadane, J. B. (1974). A characterization of triangular arrays which are expectations of order statistics. *J. Appl. Probability*, 11, 413–416. [MR 50 (1975) 15140]

99. Kotz, S. (1974). Characterizations of statistical distributions: a supplement to recent surveys. *International Statistical Review*, 42, 39–65. [MR 50 (1975) 6015]

100. Kotz, S. and Johnson, N. L. (1974). A characterization of exponential distributions by a waiting time property. *Commun. Statist.*, 3, 257–258. [MR 49 (1975) 10001]

101. Kirmani, S. N. U. A. and Alam, S. N. (1980). Characterization of the geometric distribution by the form of a predictor. *Commun. Statist. A*, 9, 541–547. [MR 81G (1981) 62023]

102. Krishnaji, N. (1971). Note on a characterizing property of the exponential distribution. *Ann. Math. Statist.*, 42, 361–362.

103. Laurent, A. G. (1974). On characterization of some distributions by truncation properties. *J. Amer. Statist. Assoc.*, 69, 823–827. [MR 52 (1976) 9441]

104. Lukacs, E. (1977). Stability theorems. *Adv. Appl. Probab.*, 9, 336–361. [MR 56 (1978) 13312]

105. MacGillivray, H. L. (1981). A note on the normalized moments of distributions with nonmonotonic hazard rate. *J.Appl. Probab.*, 18, 530–535. [MR 82M (1982) 62227]

106. Mallows, C. L. (1973). Bounds on distribution functions in terms of expectations of order statistics. *Ann. of Probab.*, 1, 297–303. [MR 50 (1975) 8828]

107. Marsaglia, G. and Tubilla, A. (1975). A note on the "lack of memory" property of the exponential distribution. *Ann. Probab.*, 3, 353–354. [MR 51 (1976) 2073]

108. Marshall, A. W. (1975). Some comments on the hazard gradient. *Stoch. Process. Appl.*, 3, 293–300. [MR 53 (1977) 4431]

109. Marshall, A. W. and Olkin, I. (1967). A multivariate exponential distribution. *J. Amer. Statist. Assoc.*, 62, 30–44. [MR 35 (1968) 6241]

110. Marshall, A. W. and Olkin, I. (1967). A generalized bivariate exponential distribution. *J. Appl. Probab.*, 4, 291–302. [MR 35 (1968) 4962]

111. Nagaraja, H. N. (1977). On a characterization based on record values. *Australian J. Statist.*, 19, 70–73. [MR 56 (1978) 16867]

112. Obretenov, A. (1977). Convergence of IFR-distribution to the exponential one. *Comptes Rendus de l'Académie Bulgare des Sciences*, 30, 1385–1387. [MR 57 (1979) 17780]

113. Paulson, A. S. (1973). A characterization of the exponential distribution and a bivariate exponential distribution. *Sankhyā*, A35, 69–78. [MR 50 (1975) 15062]

114. Pollak, M. (1973). On equal distributions. *Ann. Statist.*, 1, 180–182. [MR 48 (1974) 9914]

115. Puri, P. S. (1973). On a property of exponential and geometric distributions and its relevance to multivariate failure rate. *Sankhyā*, A35, 61–68.

116. Puri, P. S. and Rubin, H. (1970). A characterization based on the absolute difference of two i.i.d. random variables. *Ann. Math. Statist.*, 41, 2113–2122. [MR 45 (1973) 2836]

117. Puri, P. S. and Rubin, H. (1974). On a characterization of the family of distributions with constant multivariate failure rates. *Ann. Probab.*, 2, 738–740. [MR 55 (1978) 9409]

118. Ramachandran, B. (1979). On the "strong memoryless property" of the exponential and geometric probability laws. *Sankhyā*, A41, 244--251. [MR 82H (1982) 62031]

119. Reinhardt, H. E. (1968). Characterizing the exponential distribution. *Biometrics*, 24, 437–439.

120. Resnick, S. I. (1973). Limit laws for record values. *Stochast. Process. Appl.*, 1, 67–82. [MR 50 (1975) 14895]

121. Riedel, M. (1981). On Bosch's characterization of the exponential distribution function. *Zeitschrift für Angewandte Mathematik und Mechanik*, 61, 272–273.

122. Rogers, G. S. (1959). A note on the stochastic independence of functions of order statistics. *Ann. Math. Statist.*, 30, 1263–1264. [MR 22 (1961) 290]

123. Rossberg, H. J. (1972). Characterization of the exponential and the Pareto distributions by means of some properties of the distribution. *Math. Operationsforschung und Statistic*, 3, 207–216. [MR 48 (1974) 9915]

124. Rossberg, H. J. (1972). Characterization of distribution functions by the independence of certain functions of order statistics. *Sankhyā*, A34, 111–120. [MR 51 (1976) 14342]

125. Sen, P. K. (1959). On the moments of the sample quantiles. *Calcutta Statist. Ass. Bull.*, 9, 1–19. [MR 22 (1961) 288]

126. Shanbhag, D. N. (1970). The characterizations for exponential and geometric distributions. *J. Amer. Statist. Assoc.*, 65, 1256–1259.

127. Shanbhag, D. N. (1977). An extension of the Rao-Rubin characterization of the Poisson distribution. *J. Appl. Probab.*, 14, 640–646. [MR 56 (1978) 9769]

128. Shantaram, R. and Harkness, W. (1972). On a certain class of limit distributions. *Ann. Math. Statist.*, 43, 2067–2071. [MR 50 (1975) 5901]

129. Shimizu, R. (1978). Solution to a functional equation and its application to some characterization problems. *Sankhyā*, A40, 319– 332. [MR 81J (1981) 60021]

130. Shimizu, R. (1979). On a lack of memory property of the exponential distribution. *Ann. Inst. Statist. Math.*, 31, 309–313. [MR 81F (1981) 62026]

131. Shimizu, R. (1979). A characterization of the exponential distribution. *Ann. Inst. Statist. Math.*, 31, 367–372. [MR 81J (1981) 62034]

132. Shimizu, R. (1980). Functional equation with an error term and the stability of some characterizations of the exponential distribution. *Ann. Inst. Statist. Math.*, 32, 1–16. [MR 81F (1981) 62027]

133. Shimizu, R. and Davies, L. (1980). General characterization theorems for the Weibull and the stable distributions. *The Institute of Statistical Mathematics*, Research Memorandom No. 173, 35 p.

134. Shorrock, R. W. (1972). A limit theorem for inter-record times. *J. Appl. Probab.*, 9, 219–223. [MR 45 (1973) 4487]

135. Srivastava, M. S. (1967). A characterization of the exponential distribution. *Amer. Math. Monthly*, 74, 414–416. [MR 35 (1968) 7374]

136. Srivastava, R. C. (1974). Two characterizations of the geometric distribution. *J. Amer. Statist. Assoc.*, 69, 267–269. [MR 52 (1976) 1971]

137. Szantai, R. (1971). On limiting distributions for the sums of random number of random variables concerning the rarefaction of recurrent process. *Studia Sci. Math. Hungarica*, 6, 443–452.

138. Talwalker, S. (1977). A note on characterization by the conditional expectation. *Metrika*, 24, 129–136. [MR 56 (1978) 16868]

139. Tanis, E. A. (1964). Linear forms in the order statistics from an exponential distribution. *Ann. Math. Statist.*, 35, 270–276. [MR 28 (1964) 1704]

140. Tata, M. N. (1969). On outstanding values in a sequence of random variables. *Zeitschrift für Wahrscheinlichkeitstheorie und Verwandte*, 12, 9–20. [MR 40 (1970) 918]

141. Teicher, H. (1961). Maximum likelihood characterization of distributions. *Ann. Math. Statist.*, 32, 1214–1222. [MR 24A (1962) 586]

142. Westcott, M. (1981). Characterizing the exponential distribution *J. Appl. Probab.*, 18, 568.

143. Young, G. S. (1958). The linear functional equation. *Amer. Math. Monthly*, 65, 37–38. [MR 20 (1959) 4106]

AUTHOR INDEX

Aczél, J. 48, 115
Ahsanullah, M. 2, 115
Alam, S. N. 116
Ali, M. M. 115
Arnold, B. C. 115, 116
Azlarov, T. A. 113, 114, 116

Balkema, A. A. 113
Barlow, R. E. 1, 18, 23
Basu, A. P. 115, 116
Beg, M. I. 115
Bellman, R. 114
Berg, S. 116
Berk, R. H. 115
Bildikar, S. 116
Block, H. W. 116
Bolger, E. M. 115
Bosch, K. 116
Brandofe, T. 118

Cauchy, A. L. 113
Chan, L. K. 115
Chong, K. M. 115
Crawford, G. B. 116
Cundy, H. 114

Džamirzaev, A. A. 3, 113, 114, 116
Dallas, A. C. 114, 116
Darboux, G. 113
David, H. A. 24, 29, 114
Davies, L. 118

Davis, D. J. 1
Derjagin, Ju. V. 114
Desu, M. M. 115
Dimitrov, B. 12, 91
Downton, F. 116
Durrett, R. T. 113

Egorov, V. A. 114
El-Neweihi, Emad, 116
Epstein, B. 1
Erickson, K. B. 114
Esary, J. D. 116

Feller, W., 44, 116
Ferguson, T. S. 2, 114, 116
Fisz, M. 114
Freund, J. E. 116

Galambos, J. 2, 3, 114, 115, 116, 117
Gešev, A. A. 114, 115
Ghurye, S. G. 113
Gohberg, I. C. 114
Gokhale, D. V. 116
Govindarajulu, Z. 115, 116
Greenberg, B. G. 1, 114
Grosswald, E. 115
Guess, H. A. 114
Gumbel, E. J. 86, 116
Gupta,R. C. 113

Haan, L. de 113
Hamdan, M. A. 115

135

Harkness, W. L. 113, 115
Hawkins, D., 113
Heyde, C. C. 113
Hoang, Huu Nhu 113
Huang, S. T. 114, 115, 116
Hwang, J. S. 115

Isaacson, D. 115
Ismatullaev, Š. A. 3

Januškjavichene, O. 37
Johnson, N. L. 86, 113, 115, 116

Kabe, D. G. 116
Kadane, J. B. 115
Kagan, A. M. 2, 3
Karlin, S. 114
Kirmani, S. N. U. A. 115, 116
Klebanov, L. B 2, 3, 12, 37, 91, 114, 116
Kotz, S. 2, 3, 86, 113, 114, 115, 116, 117
Krein, M. G. 114
Krishnaji, N. 113

Laurent, A. J. 114
Leslie, J. R. 113
Linnik, Yu. V. 2, 3
Lukacs, E. 115

MacGillivray, H. L. 113
Mallows, C. L. 115
Marsaglia, G. 113
Marshall, A. W. 23, 86, 87, 116, 117
Meeden, G. 115

Melamed, I. A. 116
Mkrtcjan, S. T. 63, 115

Nagaraja, H. N. 115

Obretenov, A. 3, 23, 93, 113, 116
Olkin, I. 86, 87, 117

Patil, G. P. 116
Paulson, A. S. 116
Polesickaja, N. A. 114
Pollak, M. 115
Proschan, F. 1, 15, 18, 23
Puri, P. S. 116

Rachev, S. 3, 12, 91, 93
Ramachandran, B. 113, 118
Rao, C. R. 2, 3
Reinhardt, H. E. 114
Resnick, S. I. 114
Riedel, M. 116
Rogers, G. S. 114
Rossberg, H. J. 2, 114
Rubin, H. 116

Sakhabov, O. M. 114, 115
Saleh, A. K. Md. E. 115
Sarhan, A. E. 1, 114
Sen, P. K. 114
Shanbhag, D. N. 114, 116, 118
Shantaram, R. 113
Shimizu, R. 2, 34, 113, 114, 115, 118
Shorrock, R. W. 114
Siraždinov,S. H. 3

Sobel, M. 1
Solovév, A. D. 113
Srivastava, M. S. 2, 114
Srivastava, R. C. 116
Sultanova, M. M. 113, 114, 116
Szantai, R. 116

Talwalker, S. 115
Tanis, E. A. 114

Tata, M. N. 114, 115
Teicher, H. 2
Tubilla, A. 113

Westcott, M. 115

Volodin, N. A. 113

Young, G. S. 113